Challenger: Tragedy and Triumph - Unraveling the Space Shuttle Challenger Explosion

Oliver Lancaster

Published by Oliver Lancaster, 2023.

While every precaution has been taken in the preparation of this book, the publisher assumes no responsibility for errors or omissions, or for damages resulting from the use of the information contained herein.

CHALLENGER: TRAGEDY AND TRIUMPH - UNRAVELING THE SPACE SHUTTLE CHALLENGER EXPLOSION

First edition. July 31, 2023.

ISBN: 979-8223051442

Written by Oliver Lancaster.

Also by Oliver Lancaster

Watch for more at https://tinyurl.com/olanc.

Sign up to my free newsletter to get updates on new releases, FREE teaser chapters to upcoming releases and FREE digital short stories.

Or visit https://tinyurl.com/olanc

I never spam and you can unsubscribe at any time.

OLIVER LANCASTER

Disclaimer

The contents of this book, "Challenger: Tragedy and Triumph - Unraveling the Space Shuttle Challenger Explosion," are based on thorough research and reliable sources available. While utmost care has been taken to ensure accuracy, the author and publisher disclaim any liability for errors, omissions, or any consequences arising from the use of the information presented herein. Readers are encouraged to verify information with up-to-date sources and consult experts for specific inquiries related to space exploration or historical events.

Challenger: Tragedy and Triumph - Unraveling the Space Shuttle Challenger Explosion

CHALLENGER: TRAGEDY AND TRIUMPH - UNRAVELING THE SPACE SHUTTLE CHALLENGER EXPLOSION

Chapter 18: Challenger's Controversies and Conspiracy Theories

Chapter 19: The Challenger Memory in Pop Culture

Chapter 20: Looking Back - Reflections and Insights

OLIVER LANCASTER

Chapter 1: Introduction

The Space Shuttle Challenger was the second orbiter of NASA's Space Shuttle program and an iconic spacecraft that played a crucial role in shaping the history of space exploration. Designed and built by the aerospace manufacturer Rockwell International, the Challenger was named after the British naval research vessel HMS Challenger, which conducted groundbreaking scientific investigations in the late 1800s.

The Challenger orbiter was part of NASA's fleet of reusable space shuttles, which aimed to revolutionize space travel by offering a cost-effective and reusable means of reaching orbit. Its maiden flight took place on April 4, 1983, marking a significant milestone in human spaceflight. The reusable shuttle system was an ambitious endeavor, and the Challenger quickly became a symbol of NASA's ingenuity and determination to explore the cosmos.

The orbiter consisted of three main components: the orbiter vehicle, the external fuel tank, and the solid rocket boosters. The orbiter itself was a winged spacecraft with a length of 122.17 feet (37.24 meters) and a wingspan of 78.06 feet (23.79 meters). It was capable of carrying a crew of up to seven astronauts and was equipped to deploy satellites, conduct scientific experiments, and perform various missions.

Throughout its operational history, the Challenger carried out several notable missions that significantly contributed to scientific research, technology development, and international cooperation. One of its most significant achievements was the deployment of the Hubble Space Telescope during the STS-31 mission in April 1990. The Hubble Space Telescope has since become one of the most important tools in astronomy, providing unprecedented views of distant galaxies and cosmic phenomena.

Additionally, the Challenger played a pivotal role in advancing the understanding of space-based research. It hosted numerous scientific experiments in its cargo bay, covering fields such as astronomy, meteorology, biology, and materials science. These missions paved the way for future space exploration and fostered international collaboration in space research.

However, despite its achievements, the Challenger was tragically remembered for the disaster that occurred on January 28, 1986, during its tenth mission, designated STS-51-L. Approximately 73 seconds after liftoff, the Challenger exploded in a catastrophic event that resulted in the loss of all seven crew members, including Christa McAuliffe, who was set to become the first private citizen and teacher in space.

The disaster had a profound impact on NASA, the space program, and the nation as a whole. It prompted a thorough investigation into the cause of the explosion and led to significant changes in the management and safety procedures of future space missions. The Challenger tragedy became a somber reminder of the risks and challenges inherent in human spaceflight.

Despite the tragedy, the Space Shuttle Challenger's legacy is one of both triumph and tragedy. Its missions provided invaluable contributions to scientific knowledge and technological advancement, leaving an indelible mark on space exploration history. The investigation and subsequent improvements made in the wake of the disaster have contributed to increased safety and have shaped the way we approach space exploration today. In unraveling the events surrounding the Challenger explosion, we gain a deeper understanding of the complexity and risks involved in exploring the final frontier.

On the fateful day of January 28, 1986, the Space Shuttle Challenger's STS-51-L mission carried a diverse and accomplished crew of seven astronauts, each with unique expertise and contributions to the scientific and educational goals of the mission. Tragically, they lost their lives in the explosion that occurred just 73 seconds after liftoff. Let us honor their memory by introducing each crew member and acknowledging their significant contributions to the mission:

1. Francis R. Scobee (Commander):

Francis Richard Scobee, a veteran astronaut, served as the Commander of the Challenger mission. He had previously flown as the pilot of the STS-41-C mission and was selected to command the STS-51-L mission. Scobee had a distinguished career in the U.S. Air Force, where he flew more than 45 combat missions during the Vietnam War. As the mission commander, he was responsible for overseeing all aspects of the flight and ensuring the safety and success of the crew.

2. Michael J. Smith (Pilot):

Lieutenant Colonel Michael John Smith was the pilot of the Challenger mission. A seasoned naval aviator, he had logged over 4,500 hours of flight time in various aircraft. Smith's responsibilities included assisting the commander in piloting the shuttle and managing its systems during launch, orbit, and landing. Additionally, he was a skilled photographer and had planned to capture stunning images of Earth from space during the mission.

3. Ronald McNair (Mission Specialist):

Dr. Ronald Erwin McNair was the mission specialist on board the Challenger. An accomplished physicist and accomplished saxophonist, McNair had a Ph.D. in Physics from the Massachusetts Institute of Technology (MIT). He became the second African American to fly in space during his previous mission, STS-41-B. On STS-51-L, McNair was tasked with operating the remote manipulator arm to deploy a satellite.

4. Ellison Onizuka (Mission Specialist):

Colonel Ellison Shoji Onizuka was a mission specialist and the first Asian American astronaut to fly in space. As an experienced Air Force test pilot, he brought valuable expertise to the mission. Onizuka's responsibilities included overseeing the operation of various payloads and experiments on the

Challenger. He was committed to promoting diversity and excellence in the astronaut corps.

5. Judith A. Resnik (Mission Specialist):

Dr. Judith Arlene Resnik was a mission specialist and an accomplished electrical engineer. She had a Ph.D. in engineering from the University of Maryland and was a talented pianist. Resnik had previously flown on the maiden voyage of the orbiter Discovery, STS-41-D. On STS-51-L, she was involved in operating the shuttle's robotic arm and conducting experiments related to materials science.

6. Gregory Jarvis (Payload Specialist):

Gregory Bruce Jarvis was a payload specialist selected to represent Hughes Aircraft Company. As an engineer, he was responsible for conducting various experiments and observations on behalf of his employer. Jarvis' presence on the mission highlighted the increasing involvement of private companies in space missions.

7. Christa McAuliffe (Payload Specialist - Teacher in Space):

Christa McAuliffe was a civilian payload specialist and a high school social studies teacher from Concord, New Hampshire. Selected from among thousands of applicants, she was to be the first private citizen and teacher in space. As part of NASA's "Teacher in Space" program, McAuliffe had planned to conduct educational activities and live broadcasts from space, inspiring students and educators worldwide.

Each of these astronauts was selected for their expertise, dedication, and commitment to advancing human knowledge and inspiring future generations. The loss of these brave individuals in the Challenger explosion was a profound tragedy that shook the nation and the world, but their spirit of exploration and discovery continues to inspire the pursuit of knowledge and the dream of reaching for the stars.

CHALLENGER: TRAGEDY AND TRIUMPH - UNRAVELING THE SPACE SHUTTLE CHALLENGER EXPLOSION

Chapter 2: The Vision of Space Shuttle Program

The Space Shuttle program, conceived in the early 1970s, emerged from the desire to create a more cost-effective, reusable, and versatile spacecraft that could revolutionize human spaceflight. Prior to the development of the Space Shuttle, space missions were conducted using single-use rockets, which incurred exorbitant costs and limited the frequency of launches. NASA's vision was to build a spacecraft capable of regular launches, carrying both astronauts and payloads into space, while also providing a means of safe return to Earth.

Origins of the Space Shuttle Program:

The roots of the Space Shuttle program can be traced back to various studies and concepts developed during the 1960s. Notably, NASA's Space Task Group, established in 1958, laid the groundwork for human spaceflight through the Mercury and Gemini programs. These early missions demonstrated the feasibility of space travel but were limited in their capacity to carry significant payloads and return safely to Earth.

In the early 1970s, NASA proposed the Space Shuttle as the next step in space exploration. The idea of a reusable spaceplane gained traction, allowing for a more practical approach to space travel. This involved launching vertically like a rocket, conducting missions in space, and then gliding back to Earth for a horizontal landing, akin to an airplane.

Goals of the Space Shuttle Program:

1. Reusability and Cost Efficiency: The primary goal of the Space Shuttle program was to develop a reusable spacecraft capable of multiple flights. By reducing the expenses associated with every launch, NASA aimed to make space missions more cost-effective and accessible.

2. Versatility and Flexibility: The Space Shuttle's design allowed it to carry a variety of payloads, including satellites, scientific instruments, and modules for space stations like Skylab and the International Space Station (ISS). The shuttle's cargo bay was spacious, enabling it to deploy and retrieve satellites, conduct experiments, and host scientific research.

3. Space Station Support: A crucial aspect of the program's vision was to support the construction and maintenance of space stations. The shuttle's payload capacity and ability to transport astronauts and equipment facilitated the assembly of space habitats and laboratories in low Earth orbit.

4. Human Spaceflight and Exploration: The Space Shuttle was designed to carry astronauts into space, making it a key platform for human spaceflight missions. It played a vital role in transporting crews to and from space stations, fostering international cooperation in space exploration, and conducting scientific experiments in microgravity.

5. Satellite Deployment and Repair: The shuttle's capability to deploy, repair, and retrieve satellites in orbit was of immense value to both commercial and government entities. It enabled the deployment of communication, weather, and Earth observation satellites, as well as the servicing of malfunctioning satellites.

6. Advancements in Technology: The development of the Space Shuttle program spurred advances in various fields, including aerospace engineering, materials science, and thermal protection systems. Many of the technologies and techniques developed for the shuttle found applications in other aerospace endeavors and industries.

The Space Shuttle program marked a significant milestone in space exploration history. Its first launch on April 12, 1981, with the orbiter Columbia, marked a new era of reusable spaceflight. Over the years, the program successfully completed numerous missions, revolutionizing the way humans and payloads were transported to space. However, the tragic loss of the Challenger in 1986 and Columbia in 2003 underscored the inherent risks of space travel, leading to significant safety improvements and ultimately the retirement of the shuttle fleet in 2011.

Despite its conclusion, the Space Shuttle program remains a testament to human ingenuity and determination, serving as a vital stepping stone in the pursuit of space exploration, scientific discovery, and the exploration of the cosmos.

NASA's vision for the Space Shuttle missions was multi-faceted and encompassed a wide range of objectives that aimed to advance human space exploration, scientific research, and international cooperation. When the Space Shuttle program was conceived, NASA envisioned a versatile spacecraft that could carry out a myriad of missions, fostering innovation and expanding humanity's presence in space. The objectives of the shuttle missions can be summarized as follows:

1. Human Spaceflight and Crewed Missions:

At the core of NASA's vision was the goal of facilitating human spaceflight and conducting crewed missions to Earth's orbit and beyond. The Space Shuttle was designed to transport astronauts to and from space, supporting various activities such as satellite deployment, space station construction, and scientific research in microgravity.

2. Deployment and Servicing of Satellites:

One of the primary objectives of the Space Shuttle missions was to deploy various types of satellites into orbit. These satellites included communication satellites, weather observation satellites, Earth imaging satellites, and scientific instruments. The shuttle's cargo bay allowed for the safe transportation and precise deployment of satellites, significantly contributing to advances in telecommunications, weather forecasting, and Earth monitoring.

3. Conducting Scientific Research:

CHALLENGER: TRAGEDY AND TRIUMPH - UNRAVELING THE SPACE SHUTTLE CHALLENGER EXPLOSION

The Space Shuttle provided a unique platform for conducting scientific experiments in the microgravity environment of space. Astronauts onboard could conduct research in fields such as biology, materials science, fluid dynamics, and astronomy. The shuttle's spacious payload bay accommodated scientific instruments and laboratory facilities, making it an essential tool for researchers and scientists from around the world.

4. Assembly and Support of Space Stations:

NASA envisioned the Space Shuttle as a crucial asset in supporting the construction, maintenance, and resupply of space stations. Notably, it played a pivotal role in the assembly of Skylab, a precursor to the International Space Station (ISS). Later, during the development and utilization of the ISS, the shuttle played a vital role in transporting crews, equipment, and supplies to the orbiting laboratory.

5. International Cooperation and Diplomacy:

The Space Shuttle missions fostered international collaboration in space exploration. NASA invited astronauts from various countries to participate in shuttle missions, promoting goodwill and cooperation between nations. Notable collaborations included the inclusion of international astronauts and payload specialists from countries such as Canada, Japan, and Germany.

6. Technological Advancements and Innovation:

NASA's vision for the shuttle missions encompassed technological innovation and advancements in aerospace engineering. The challenges posed by reusable spaceflight encouraged the development of novel technologies, including thermal protection systems, reusable rocket engines, and advanced avionics. Many of these technologies found applications beyond the Space Shuttle program and continue to influence the aerospace industry.

7. Inspiration and Education:

The inclusion of payload specialists like Christa McAuliffe, the first "Teacher in Space," exemplified NASA's commitment to inspiring the next generation of explorers and educators. The shuttle missions were a powerful tool for engaging the public, captivating students, and promoting scientific curiosity and education worldwide.

By fulfilling these objectives, the Space Shuttle program aimed to further humanity's understanding of space, contribute to technological progress, and pave the way for future endeavors in human space exploration. Although the program faced challenges and experienced tragic setbacks, its legacy remains significant, leaving an enduring impact on space exploration and shaping the path for future missions beyond Earth's boundaries.

CHALLENGER: TRAGEDY AND TRIUMPH - UNRAVELING THE SPACE SHUTTLE CHALLENGER EXPLOSION

Chapter 3: The Challenger Mission (STS-51-L)

───

The STS-51-L mission of the Space Shuttle Challenger was intended to be the twenty-fifth flight of NASA's Space Shuttle program. The mission was scheduled for launch on January 22, 1986, but faced multiple delays due to technical and weather-related issues. Ultimately, the mission's ill-fated launch occurred on January 28, 1986, and tragically ended in disaster.

Background and Crew Selection:

The STS-51-L mission was highly anticipated and widely publicized due to the inclusion of the first private citizen and teacher in space. NASA's "Teacher in Space" program aimed to promote science education and inspire students across the United States by sending a teacher on a space mission. The program garnered significant interest, attracting thousands of applicants from educators across the country.

Christa McAuliffe, a high school social studies teacher from Concord, New Hampshire, was selected as the "Teacher in Space" from a competitive pool of applicants. Her passion for education, enthusiasm for space exploration, and her ability to connect with students made her a natural fit for this groundbreaking opportunity. Christa's selection as a payload specialist for the STS-51-L mission captured the nation's attention, making her a symbol of hope and aspiration for educators and students alike.

Mission Objectives:

The primary objective of the STS-51-L mission was to deploy the Tracking and Data Relay Satellite (TDRS)-B, the second satellite in the TDRS constellation. TDRS satellites were vital for ensuring continuous and reliable communication between spacecraft in orbit and ground stations on Earth. They played a crucial role in supporting various NASA missions, including space shuttle flights and space science missions.

Additionally, the STS-51-L mission aimed to conduct various scientific experiments and observations, including the study of the Earth's atmosphere and the behavior of materials in microgravity. The orbiter carried several experiments and instruments in its cargo bay, highlighting the shuttle's versatility as a platform for scientific research.

The mission also included plans for educational activities led by Christa McAuliffe. She had prepared a series of educational lessons and demonstrations to be conducted from space, allowing students to participate in the excitement of space exploration and connect with the crew in real-time. Christa's presence on the mission was intended to ignite students' interest in science, technology, engineering, and mathematics (STEM) fields and serve as a catalyst for educational outreach.

The Tragic Disaster:

CHALLENGER: TRAGEDY AND TRIUMPH - UNRAVELING THE SPACE SHUTTLE CHALLENGER EXPLOSION

Despite numerous delays, the STS-51-L mission was given the green light for launch on the morning of January 28, 1986. Just 73 seconds after liftoff, disaster struck as the Challenger orbiter disintegrated in a devastating explosion. The explosion resulted from the failure of an O-ring seal in one of the solid rocket boosters, which allowed hot gases to breach the joint and ignite the external fuel tank.

The tragic event claimed the lives of all seven crew members on board, including Christa McAuliffe, Francis R. Scobee, Michael J. Smith, Ronald McNair, Ellison Onizuka, Judith A. Resnik, and Gregory Jarvis. The nation and the world were shocked and deeply saddened by the loss, and the shuttle program was immediately suspended pending a thorough investigation into the cause of the explosion.

The disaster had a profound impact on the space program, leading to comprehensive safety improvements and a reevaluation of NASA's approach to human spaceflight. The crew of STS-51-L will always be remembered for their bravery and dedication to the pursuit of knowledge, and their memory remains an enduring part of the history of space exploration.

The crew of the ill-fated STS-51-L mission, aboard the Space Shuttle Challenger, underwent extensive training and preparation to ensure they were ready for the complexities and challenges of spaceflight. The training process for a shuttle mission was rigorous and multifaceted, encompassing a wide range of disciplines and scenarios to equip the astronauts with the necessary skills and knowledge for their mission.

OLIVER LANCASTER

1. Basic Astronaut Training:

Before specific mission training, each astronaut underwent a standard set of basic astronaut training at NASA's Johnson Space Center in Houston, Texas. This training included lessons in space systems, spacecraft operations, flight dynamics, and emergency procedures. They also learned about the shuttle's systems, simulators, and the intricacies of life aboard the spacecraft.

2. Mission-Specific Training:

Following basic training, the STS-51-L crew members engaged in mission-specific training tailored to the objectives and payload of their flight. Training modules included simulators that replicated shuttle operations, where astronauts practiced launch, rendezvous, and landing procedures.

3. Payload Operations:

As the mission aimed to deploy the Tracking and Data Relay Satellite (TDRS)-B and conduct scientific experiments, crew members received specialized training in handling the satellite and conducting scientific research in microgravity. They practiced deploying and retrieving payloads in a simulated space environment.

4. Simulated Scenarios:

Astronauts participated in a variety of simulated scenarios to prepare for potential emergencies during the flight. These simulations included launch abort procedures, dealing with onboard malfunctions, and managing critical situations during launch, orbit, and re-entry.

5. Role-Specific Training:

Each crew member had specific roles and responsibilities based on their expertise. The mission commander, Francis R. Scobee, received training on shuttle systems and flight procedures, while pilot Michael J. Smith focused on shuttle operations and navigation. Mission specialists, Ronald McNair, Ellison Onizuka, Judith A. Resnik, and payload specialist Gregory Jarvis, underwent training specific to their scientific experiments and payload operations.

6. Teacher in Space Training:

Christa McAuliffe, the first private citizen and teacher in space, underwent additional training to prepare for her role as the payload specialist. She participated in an intensive program known as the "Teacher in Space" training, which included astronaut training components along with education and outreach preparation. This training aimed to help her conduct educational activities and engage students during the mission.

7. Team Building and Communication:

Effective teamwork and communication were essential for the success of the mission. The crew members participated in team-building exercises to foster strong bonds and ensure seamless collaboration during the mission.

8. Pre-Flight Simulations:

In the weeks leading up to the scheduled launch, the crew participated in full-scale simulations, known as the Flight Readiness Review, where they conducted a dress rehearsal of the mission from launch to landing. These simulations involved mission control teams and ground support personnel to ensure coordination between on-orbit and ground-based operations.

Despite the extensive preparation and training, the STS-51-L mission tragically ended in disaster due to the explosion just after liftoff. The dedication and professionalism of the crew during their training and preparations remain a testament to their commitment to space exploration and advancing the frontiers of knowledge, and their memory continues to inspire the pursuit of exploration and discovery.

CHALLENGER: TRAGEDY AND TRIUMPH - UNRAVELING THE SPACE SHUTTLE CHALLENGER EXPLOSION

Chapter 4: The Fateful Day: Launch and Explosion

———

The launch of the Space Shuttle Challenger for the STS-51-L mission was a culmination of months of preparation and readiness reviews. However, various factors and decisions preceding the launch ultimately contributed to the tragic disaster that unfolded on January 28, 1986.

1. Mission Scheduling and Delays:

The STS-51-L mission was initially scheduled for launch on January 22, 1986. However, technical and weather-related issues caused a series of delays. The launch was postponed several times due to problems with the shuttle's main engines and weather conditions at the Kennedy Space Center in Florida.

2. Cold Weather Concerns:

On the morning of January 28, 1986, the launch temperature was unusually cold, with overnight temperatures dropping significantly below freezing. The cold weather raised concerns among engineers and managers about the impact on the performance of the shuttle's O-ring seals, which were critical components in the solid rocket boosters.

3. O-Ring Concerns and Engineering Communication:

The O-rings were designed to seal joints in the solid rocket boosters and prevent the escape of hot gases during ignition. NASA engineers had long been concerned about the reliability of the O-rings, especially in low temperatures. Prior launches had shown evidence of O-ring erosion, indicating that the seals were not as reliable as initially thought.

Engineers from Morton Thiokol, the contractor responsible for the solid rocket boosters, had expressed their concerns about the O-rings in the days leading up to the launch. They recommended postponing the launch due to the unusually cold weather conditions. However, due to significant schedule pressures and management decisions, NASA officials proceeded with the launch despite the objections raised by the engineers.

4. Pre-Launch Preparations:

On the morning of January 28, 1986, the Challenger's crew arrived at the Kennedy Space Center, completing their final pre-launch preparations. The crew members were Commander Francis R. Scobee, Pilot Michael J. Smith, Mission Specialists Ronald McNair, Ellison Onizuka, Judith A. Resnik, Payload Specialist Gregory Jarvis, and Payload Specialist and "Teacher in Space" Christa McAuliffe.

5. The Launch:

The countdown proceeded as planned, and at 11:38 AM EST, the Challenger began its ascent into space. Approximately 73 seconds after liftoff, a catastrophic explosion occurred. The explosion was triggered by the failure of the O-ring seals in one of the solid rocket boosters, allowing hot gases to escape and ignite the external fuel tank. The shuttle disintegrated, resulting in the loss of all seven crew members on board.

6. Aftermath and Investigation:

The tragic disaster of the Challenger shocked the nation and the world. The explosion led to an immediate suspension of the Space Shuttle program, and a thorough investigation was launched to determine the cause of the accident. The investigation, known as the Rogers Commission, revealed the crucial role of the O-ring seals in the disaster and exposed systemic flaws in communication and decision-making within NASA.

7. Impact and Lessons Learned:

The Challenger disaster resulted in a significant reassessment of safety protocols and procedures within NASA. It led to the implementation of comprehensive changes in the shuttle program, including the redesign of the solid rocket boosters, improvements in communication between engineers and management, and enhanced safety measures for future missions.

The events leading up to the launch of the Challenger highlighted the importance of prioritizing safety over schedule pressures and the significance of open communication in decision-making. The crew members of STS-51-L will always be remembered for their dedication and bravery, and the tragedy remains a solemn reminder of the inherent risks and challenges associated with space exploration.

The explosion of the Space Shuttle Challenger during the STS-51-L mission was a result of a combination of technical and organizational factors. A critical failure in the O-ring seals of the solid rocket boosters, coupled with management decisions, communication breakdowns, and schedule pressures, ultimately led to the tragic disaster. Let's analyze the key factors that contributed to the explosion and subsequent loss of the Challenger and its crew:

1. O-Ring Seal Design and Performance:

The primary technical factor behind the disaster was the design and performance of the O-ring seals used in the solid rocket boosters. The O-rings were meant to prevent hot gases from escaping through joints between segments of the boosters during ignition. However, they were found to be vulnerable to failure, particularly in low temperatures.

On the morning of the launch, the temperature at Kennedy Space Center was unusually cold, well below freezing. The cold weather made the O-rings less resilient and more susceptible to failure. This was evident in previous launches where O-ring erosion was observed. Despite these concerns, the decision was made to proceed with the launch, underestimating the risks posed by the cold weather to the integrity of the O-ring seals.

2. Communication Breakdowns:

The Rogers Commission, which investigated the disaster, identified serious communication breakdowns between NASA's engineers and decision-makers. Engineers from Morton Thiokol, the contractor responsible for the solid rocket boosters, raised concerns about the O-rings and recommended delaying the launch. However, their concerns were not adequately conveyed to top-level NASA officials, who were under pressure to maintain the launch schedule.

The decision-making process lacked transparency, and the critical information regarding the O-ring issues was not effectively communicated to key decision-makers. As a result, NASA officials did not fully understand the severity of the O-ring problem and proceeded with the launch despite the warnings.

3. Organizational Pressures and Schedule Constraints:

NASA was facing significant schedule pressures in the lead-up to the launch. The Challenger's mission had already been delayed multiple times, and there was a desire to keep to the launch schedule. There was also immense public and media interest due to the presence of Christa McAuliffe, the "Teacher in Space," which added to the pressure to proceed with the mission.

The schedule constraints and the emphasis on meeting launch deadlines may have influenced the decision-making process and led to the underestimation of the risks associated with the O-ring issues.

4. Safety Culture and Organizational Issues:

The disaster also revealed deeper organizational and safety culture issues within NASA. The space agency's emphasis on the success of the shuttle program and its public image may have overshadowed a more cautious approach to risk management. There was a lack of a safety culture that encouraged open discussion of concerns and a willingness to prioritize safety over schedule.

The Challenger disaster prompted NASA to reevaluate its safety protocols, management practices, and communication channels to prevent similar tragedies in the future.

The explosion of the Challenger and the subsequent loss of its crew were the result of a complex interplay of technical flaws, communication breakdowns, organizational pressures, and safety culture issues. The failure of the O-ring seals due to cold weather, combined with miscommunication and a drive to meet launch schedules, led to a catastrophic outcome that shook the foundations of the Space Shuttle program and forever altered the course of human space exploration. The disaster served as a stark reminder of the importance of prioritizing safety, fostering open communication, and learning from failures in the pursuit of space exploration.

The technical details surrounding the O-rings and other components that failed during the Space Shuttle Challenger's STS-51-L mission are essential to understanding the cause of the tragic disaster. Let's explore these aspects in detail:

1. Solid Rocket Boosters (SRBs):

The Space Shuttle's launch system consisted of two solid rocket boosters (SRBs) attached to the sides of the external fuel tank. These boosters were responsible for providing most of the thrust during liftoff and initial ascent. Each SRB was constructed in segments that were assembled and sealed with field joints.

2. O-Ring Seals:

The SRBs used O-ring seals to prevent the escape of hot gases at the field joints during ignition. The O-rings were made of synthetic rubber and were placed in grooves between each segment to create a tight seal. The effectiveness of these O-rings was crucial in maintaining the structural integrity of the SRBs during the shuttle's ascent.

3. Joint Design:

The field joint design of the SRBs had two primary segments - the forward and aft field joints. These joints were subjected to intense pressure and stress during launch and ascent. The O-rings in these joints were the primary barrier to prevent hot gases from escaping and damaging the integrity of the SRBs.

4. O-Ring Vulnerability:

One of the critical technical flaws in the SRB design was the vulnerability of the O-rings to failure, particularly in cold weather conditions. The O-rings were more pliable at higher temperatures, allowing for a better seal. However, at low temperatures, such as those experienced during the STS-51-L launch, the O-rings became less resilient and prone to not sealing effectively.

5. Temperature on Launch Day:

On the morning of January 28, 1986, the temperature at Kennedy Space Center was 36 degrees Fahrenheit (2 degrees Celsius). This was well below the lower temperature limit for the O-rings' optimal performance. The cold weather made the O-rings stiff and less likely to expand and create a secure seal. The cold conditions significantly increased the risk of O-ring failure.

6. Challenger's Ascent:

As the Challenger lifted off, hot gases from within the solid rocket boosters began to escape through the failed O-rings at the aft field joint of the right SRB. This created a breach in the joint, leading to the leakage of hot gases into the external fuel tank area.

7. Explosion and Structural Failure:

The escaping hot gases impinged on the external fuel tank, causing structural damage and ignition of the hydrogen fuel and oxygen tank. This catastrophic chain of events led to the explosion of the external fuel tank, disintegration of the orbiter, and the tragic loss of all seven crew members on board.

8. Engineering and Communication Issues:

The Rogers Commission, which investigated the disaster, revealed that engineers from Morton Thiokol, the contractor responsible for the SRBs, had raised concerns about the O-rings and recommended delaying the launch due to the cold weather. However, communication breakdowns between the engineers and NASA decision-makers resulted in the concerns not being adequately conveyed, and the launch proceeded despite the known risk.

In the aftermath of the disaster, NASA implemented significant changes to improve the safety of the Space Shuttle program, including redesigning the solid rocket boosters, enhancing communication channels, and fostering a stronger safety culture. The Challenger disaster remains a poignant reminder of the importance of technical rigor, thorough risk assessment, and open communication in the pursuit of space exploration.

Chapter 5: Investigation and Fallout

———

Immediate Aftermath of the Explosion and NASA's Response:

1. The Tragic Explosion:

On January 28, 1986, the Space Shuttle Challenger experienced a catastrophic explosion 73 seconds after liftoff during the STS-51-L mission. The explosion led to the disintegration of the orbiter, resulting in the loss of all seven crew members on board, including Christa McAuliffe, the first "Teacher in Space." The tragedy was witnessed by millions of people watching the live broadcast and had a profound impact on the nation and the world.

2. Emergency Response and Grounding of the Shuttle Fleet:

Immediately after the explosion, NASA declared a contingency emergency and initiated an immediate response. Search and rescue teams were dispatched to the launch site to recover debris and search for survivors, though none were found. The Space Shuttle program was immediately grounded, suspending all shuttle missions until a thorough investigation into the cause of the disaster could be conducted.

3. The Rogers Commission:

President Ronald Reagan established the Presidential Commission on the Space Shuttle Challenger Accident, known as the Rogers Commission, to investigate the disaster. The commission was led by former Secretary of State William P. Rogers and included experts from various fields, including aerospace, engineering, and management.

4. Investigation and Findings:

The Rogers Commission conducted an extensive investigation to determine the cause of the explosion. Over several months, the commission gathered and analyzed evidence, including telemetry data, photographs, and witness testimonies. They identified the critical role of the O-rings in the failure of the solid rocket boosters, as well as the breakdowns in communication and decision-making processes within NASA.

5. Presentation of the Report:

On June 9, 1986, the Rogers Commission presented its findings and recommendations to President Reagan and the public. The report highlighted the technical flaws in the O-ring design, the communication breakdowns between engineers and management, and the pressure to maintain the launch schedule despite safety concerns.

6. Redesign and Safety Improvements:

NASA immediately took steps to address the issues identified by the Rogers Commission. The solid rocket boosters were redesigned to include improved O-rings and additional safety measures. The space agency implemented enhanced communication protocols to ensure that critical safety concerns would be effectively communicated to top-level decision-makers.

7. Leadership Changes:

In the wake of the disaster, NASA underwent significant leadership changes. William R. Graham was appointed as the Acting Administrator of NASA, and Lawrence B. Mulloy, the manager of the Solid Rocket Booster Project, was reassigned.

8. Return to Flight:

After a rigorous redesign and safety improvements, NASA prepared to resume shuttle missions. The first post-Challenger flight was STS-26, launched on September 29, 1988, using the Space Shuttle Discovery. This marked the successful return to spaceflight after a hiatus of more than two years.

9. Honoring the Crew:

In the aftermath of the disaster, NASA and the nation paid tribute to the fallen crew members. The agency established the Space Shuttle Challenger Center for Space Science Education, dedicated to fostering STEM education and inspiring future generations of explorers.

The Challenger disaster had a lasting impact on NASA and space exploration, leading to significant safety improvements and a heightened emphasis on risk assessment and open communication. The tragedy also reinforced the inherent risks of human spaceflight and served as a solemn reminder of the importance of safety in the pursuit of space exploration. The memory of the STS-51-L crew members remains an enduring part of space history and a symbol of bravery and dedication to the advancement of human knowledge.

The establishment of the Rogers Commission and its investigation was a crucial step in understanding the causes of the Space Shuttle Challenger disaster and determining the factors that led to the tragic explosion. Let's explore how the commission was formed and the key aspects of its investigation:

1. Formation of the Commission:

In the aftermath of the Space Shuttle Challenger explosion on January 28, 1986, President Ronald Reagan decided to create an independent commission to investigate the disaster. The commission was formally established on February 3, 1986, and was officially known as the "Presidential Commission on the Space Shuttle Challenger Accident." It became popularly known as the Rogers Commission after its chairman, former Secretary of State William P. Rogers.

2. Composition of the Commission:

The Rogers Commission was composed of experts from various fields, including aerospace, engineering, management, and public policy. In addition to William P. Rogers, other prominent members of the commission included Neil Armstrong, the first person to walk on the moon; Sally Ride, the first American woman in space; and Richard Feynman, a Nobel Prize-winning physicist.

3. Mission and Objectives:

CHALLENGER: TRAGEDY AND TRIUMPH - UNRAVELING THE SPACE SHUTTLE CHALLENGER EXPLOSION

The primary mission of the Rogers Commission was to conduct a comprehensive investigation into the Space Shuttle Challenger disaster. Its main objectives were to determine the technical causes of the explosion and to identify any contributing factors related to management, communication, and decision-making processes within NASA and its contractors.

4. Data Collection and Analysis:

The commission's investigation involved gathering an extensive array of data, including telemetry data, video footage, photographs, technical documentation, and witness testimonies. It sought to reconstruct the sequence of events leading up to the explosion and understand the performance of the solid rocket boosters, particularly the O-ring seals, during the shuttle's ascent.

5. Public Hearings and Testimonies:

The Rogers Commission held public hearings to gather information and testimonies from key individuals involved in the shuttle program, including NASA officials, engineers, and contractors. These hearings provided insights into the decision-making process and communication practices leading up to the launch.

6. Technical Analysis and Experiments:

To understand the behavior of the O-rings and other critical components, the commission conducted its own technical analysis and experiments. This included testing the O-rings under various conditions to assess their vulnerability to failure, especially in cold temperatures.

7. Identification of the Causes:

Through its investigation, the Rogers Commission identified the technical failure of the O-ring seals in the solid rocket boosters as the primary cause of the explosion. It concluded that the O-rings had not properly sealed due to the unusually cold temperatures on the morning of the launch, allowing hot gases to escape and damage the integrity of the boosters.

8. Management and Communication Issues:

In addition to the technical cause, the commission also uncovered issues related to the decision-making process and communication breakdowns within NASA and its contractors. It highlighted concerns raised by engineers from Morton Thiokol about the O-rings but found that these concerns were not adequately conveyed to top-level decision-makers, contributing to the launch proceeding despite known risks.

9. Presentation of the Report:

The Rogers Commission presented its findings and recommendations to President Ronald Reagan and the public on June 9, 1986. The report provided a comprehensive analysis of the technical and organizational factors that led to the disaster and offered recommendations to improve safety and decision-making processes within the space program.

The Rogers Commission's investigation was critical in revealing the causes of the Challenger disaster and identifying the necessary improvements to prevent such tragedies in the future. Its thorough analysis and recommendations paved the way for significant safety enhancements in the Space Shuttle program and influenced NASA's approach to space exploration for years to come.

The findings of the Rogers Commission, which investigated the Space Shuttle Challenger disaster, had a profound impact on the space program, leading to significant changes in safety practices, decision-making processes, and the overall culture within NASA. The commission's analysis exposed critical flaws and weaknesses in the space agency's operations, prompting reforms that aimed to prevent similar accidents in the future. Let's analyze the key findings of the commission and their impact on the space program:

1. Technical Cause - O-Ring Failure:

The Rogers Commission's primary finding was that the disaster was caused by the failure of the O-ring seals in the solid rocket boosters. The commission identified that the O-rings had not properly sealed due to the unusually cold temperatures on the morning of the launch. This led to hot gases escaping and damaging the integrity of the boosters, ultimately resulting in the explosion.

Impact: The technical cause of the disaster prompted NASA to undertake significant redesign efforts to improve the safety and reliability of the solid rocket boosters. The O-rings were modified to be more resilient to extreme temperatures, and additional safety measures were implemented to ensure better performance during launch.

2. Communication Breakdowns:

The commission's investigation revealed significant communication breakdowns between engineers from Morton Thiokol, the contractor responsible for the solid rocket boosters, and top-level decision-makers at NASA. Concerns about the O-rings and the launch were not effectively conveyed to those with the authority to make critical decisions.

Impact: NASA recognized the importance of open and transparent communication between all levels of the organization. The agency implemented changes to encourage engineers and personnel to speak up about safety concerns and ensure that information reaches the appropriate decision-makers in a timely manner.

3. Management Decisions:

The commission's report pointed to management decisions and the pressure to maintain the launch schedule as factors that influenced the decision to proceed with the launch despite the known risks associated with the O-rings and the cold weather.

Impact: The tragedy underscored the importance of safety over schedule and compelled NASA to reevaluate its approach to decision-making. The space agency placed greater emphasis on thorough risk assessment, weighing safety considerations above all else when determining whether to proceed with a launch.

4. Safety Culture:

The commission's investigation highlighted a lack of a strong safety culture within NASA, where concerns about safety were not given enough weight in decision-making processes.

Impact: The Challenger disaster led to a cultural shift within NASA, emphasizing safety as the top priority in all aspects of spaceflight. A stronger safety culture was cultivated, encouraging employees to voice concerns and advocate for safety measures without fear of retribution.

5. Impact on Future Missions:

The findings of the commission had a lasting impact on the future of the Space Shuttle program and NASA's approach to space exploration. Safety improvements implemented following the disaster helped enhance the reliability of the shuttle fleet, leading to a safer and more successful series of subsequent missions.

6. Public Perception and Trust:

The Challenger disaster had a profound effect on the public's perception of spaceflight and NASA. The tragedy and subsequent investigation raised concerns about the risks associated with human space exploration and the need for transparent communication from space agencies.

Impact: NASA made concerted efforts to rebuild public trust by prioritizing safety and being more forthcoming about risks and challenges associated with space missions. The agency also sought to engage the public in its space programs through educational outreach and fostering greater transparency in its operations.

The findings of the Rogers Commission following the Challenger disaster fundamentally reshaped NASA's approach to space exploration. The investigation's emphasis on safety, communication, and decision-making processes instigated a cultural transformation within the space agency. The reforms that followed the disaster greatly improved the safety and reliability of subsequent space missions, and NASA's commitment to transparency and safety has become an enduring legacy of the Challenger tragedy.

CHALLENGER: TRAGEDY AND TRIUMPH - UNRAVELING THE SPACE SHUTTLE CHALLENGER EXPLOSION

Chapter 6: Challenger Crew Tribute

In honor of the brave crew members who lost their lives aboard the Space Shuttle Challenger during the STS-51-L mission, we pay tribute to their remarkable lives, indelible legacies, and their enduring contributions to space exploration and humanity. Each member of the crew possessed unique qualities and accomplishments that continue to inspire us today:

1. Francis R. Scobee (Commander):

As the mission commander, Francis R. Scobee displayed extraordinary leadership and dedication to the pursuit of knowledge. A veteran astronaut, Scobee had previously piloted shuttle missions and served as an accomplished test pilot. His calm and steady demeanor earned him the admiration and respect of his peers, making him an exceptional leader.

2. Michael J. Smith (Pilot):

Pilot Michael J. Smith was an accomplished Navy pilot with a passion for flying. His expertise in aeronautics and his dedication to the space program made him a vital part of the STS-51-L mission. Smith's passion for exploration and adventurous spirit were emblematic of the courage it takes to venture beyond our world.

3. Ronald McNair (Mission Specialist):

Ronald McNair was a gifted physicist and accomplished saxophonist. His exceptional academic achievements and pioneering research in laser physics made him an invaluable asset to the scientific community. McNair's determination to break barriers and shatter stereotypes serves as an enduring inspiration to aspiring scientists and marginalized communities.

4. Ellison Onizuka (Mission Specialist):

Ellison Onizuka, the first Asian American astronaut, brought a wealth of experience as an aerospace engineer and Air Force test pilot to the STS-51-L mission. His determination to overcome adversity and cultural barriers exemplified the spirit of exploration and the power of diversity in space exploration.

5. Judith A. Resnik (Mission Specialist):

As a distinguished electrical engineer and aerospace engineer, Judith A. Resnik was a pioneer for women in STEM fields. Her brilliance and achievements in the male-dominated world of engineering set an example for countless women aspiring to break barriers in science and engineering.

6. Gregory Jarvis (Payload Specialist):

Gregory Jarvis, a skilled engineer and astronaut, was the payload specialist on the STS-51-L mission. His passion for scientific research and space exploration was evident in his dedication to his mission objectives. Jarvis's desire to further humanity's understanding of space and technology serves as a testament to the importance of exploration.

CHALLENGER: TRAGEDY AND TRIUMPH - UNRAVELING THE SPACE SHUTTLE CHALLENGER EXPLOSION

7. Christa McAuliffe (Payload Specialist - "Teacher in Space"):

Christa McAuliffe, the first private citizen and "Teacher in Space," embodied the spirit of education and inspiration. As an educator, she sought to ignite the love of learning and exploration in students worldwide. Her presence on the mission captured the nation's imagination and left an indelible mark on the hearts of educators and students alike.

The loss of the Challenger crew was a poignant reminder of the inherent risks in space exploration, but their sacrifice and legacy continue to reverberate through time. Each member of the crew embodied the spirit of exploration, dedication to knowledge, and the pursuit of dreams. They have left an enduring impact on space exploration, scientific research, and the power of human resilience.

As we remember the Challenger crew, we honor their courage, their passion for discovery, and their commitment to pushing the boundaries of human potential. They will forever be remembered as heroes who dared to venture into the unknown and as beacons of hope, inspiration, and the spirit of exploration that drives us to reach for the stars. May their legacy inspire generations to come to dream, explore, and strive for a brighter future beyond the confines of Earth.

1. Francis R. Scobee (Commander):

Francis R. Scobee's remarkable achievements in space exploration were a testament to his leadership and dedication. Before commanding the ill-fated STS-51-L mission, Scobee served as pilot for STS-41-C and STS-51-I missions. His experience as a test pilot and astronaut made him a highly respected figure in the space program. His leadership during STS-51-L showcased his ability to stay composed under pressure and make crucial decisions.

Impact on Space Exploration:

Scobee's contributions helped pave the way for future shuttle missions and human spaceflight endeavors. His leadership and professionalism left an indelible mark on NASA's safety protocols and the importance of thorough risk assessment. His legacy continues to inspire astronauts and space enthusiasts to pursue excellence in their careers and uphold the values of courage and determination.

2. Michael J. Smith (Pilot):

Michael J. Smith's passion for flying and his expertise in aeronautics were instrumental in his selection as a shuttle pilot. Before joining NASA, Smith served in the United States Navy and held various positions as a test pilot. He brought invaluable flight experience and dedication to the STS-51-L mission.

Impact on Space Exploration:

Smith's accomplishments as a pilot and his willingness to explore new frontiers exemplified the adventurous spirit of space exploration. His dedication to safety and the pursuit of knowledge helped shape the culture of NASA and influenced future shuttle missions. His legacy reminds us of the crucial role pilots play in space missions and serves as a reminder of the risks and rewards associated with human spaceflight.

3. Ronald McNair (Mission Specialist):

Ronald McNair was a visionary physicist and accomplished saxophonist whose achievements defied racial and social barriers. Before joining NASA, McNair earned a Ph.D. in physics, conducting groundbreaking research in laser technology. His passion for both science and the arts made him an inspiring figure.

Impact on Space Exploration:

McNair's accomplishments as a scientist and his advocacy for education and diversity continue to inspire aspiring scientists, particularly those from marginalized communities. His legacy has contributed to efforts to increase diversity in the space program and the scientific community at large. His life and achievements serve as a symbol of perseverance, breaking barriers, and the boundless potential of the human mind.

4. Ellison Onizuka (Mission Specialist):

Ellison Onizuka's accomplishments as an aerospace engineer and Air Force test pilot earned him a place as the first Asian American astronaut. His expertise in aerospace engineering and space technology made him an invaluable asset to the STS-51-L mission.

Impact on Space Exploration:

Onizuka's legacy as the first Asian American astronaut has served as an inspiration to generations of Asian Americans interested in space exploration and engineering. His achievements underscore the importance of diversity and inclusion in the space program and have motivated countless individuals to pursue careers in science, technology, engineering, and mathematics (STEM) fields.

5. Judith A. Resnik (Mission Specialist):

Judith A. Resnik was a pioneering electrical engineer and aerospace engineer, breaking barriers for women in STEM fields. Before joining NASA, Resnik made significant contributions to engineering research and technology.

Impact on Space Exploration:

Resnik's trailblazing achievements as a woman in engineering paved the way for future generations of female scientists and engineers. Her legacy serves as a reminder of the importance of gender equality in the space industry and the value of diverse perspectives in solving complex challenges. Resnik's pioneering spirit continues to inspire young women to pursue careers in STEM and embrace their potential to excel in traditionally male-dominated fields.

6. Gregory Jarvis (Payload Specialist):

Gregory Jarvis was a skilled engineer whose passion for scientific research and space exploration made him an ideal payload specialist for the STS-51-L mission. Before joining NASA, Jarvis worked on various engineering projects, showcasing his expertise in the field.

Impact on Space Exploration:

Jarvis's dedication to advancing scientific research in space serves as a reminder of the critical role of payload specialists in conducting experiments and investigations aboard the shuttle. His legacy highlights the importance of scientific curiosity and exploration in space and has encouraged ongoing research to benefit humanity both on Earth and in space.

7. Christa McAuliffe (Payload Specialist - "Teacher in Space"):

Christa McAuliffe's selection as the "Teacher in Space" marked a groundbreaking moment in space exploration. As a high school teacher with a passion for education, McAuliffe's presence on the mission aimed to inspire students and educators worldwide.

Impact on Space Exploration:

McAuliffe's legacy lies in her dedication to promoting education and inspiring the next generation of scientists, engineers, and explorers. Her willingness to represent teachers and educators in space underscored the power of education to ignite curiosity and passion for space and science. Though her life was tragically cut short, her spirit lives on in educational outreach programs that continue to inspire students to dream big and reach for the stars.

The crew members of the Space Shuttle Challenger's STS-51-L mission left behind an enduring legacy of courage, dedication, and contributions to space exploration. Their individual achievements and sacrifices have inspired countless individuals to pursue careers in science, engineering, education, and aerospace. Their impact on space exploration and humanity's understanding of the cosmos continues to be felt, and their memories serve as a reminder of the risks and rewards of venturing into the great unknown. Their names and achievements will forever be inscribed in the annals of space history, as their spirit lives on in the pursuit of knowledge and the exploration of the cosmos.

CHALLENGER: TRAGEDY AND TRIUMPH - UNRAVELING THE SPACE SHUTTLE CHALLENGER EXPLOSION

Chapter 7: Lessons Learned and Safety Reforms

———

The Space Shuttle Challenger disaster on January 28, 1986, was a tragic event that deeply impacted NASA and the entire space community. The disaster led to profound lessons being learned that have since shaped the approach to space exploration and safety. Here are some of the key lessons learned from the Challenger disaster:

1. Safety Must Be Paramount:

One of the most significant lessons from the Challenger disaster was the importance of prioritizing safety above all else. The tragedy served as a stark reminder that space exploration is inherently risky, and every effort must be made to mitigate those risks to ensure the safety of astronauts and the success of missions.

2. Thorough Risk Assessment:

The Challenger disaster exposed the need for thorough risk assessment and analysis during the planning and preparation of space missions. Understanding and addressing potential risks, even those that may seem unlikely, is critical to prevent catastrophic failures.

3. Open and Transparent Communication:

Effective communication is vital in any organization, especially in endeavors as complex as space exploration. The disaster highlighted the need for open and transparent communication channels, ensuring that all relevant information, concerns, and risks are effectively conveyed to decision-makers and across all levels of the organization.

4. Redundancy and Contingency Planning:

The Challenger disaster underscored the importance of redundancy in critical systems and the need for robust contingency planning. Having backup systems and procedures can be vital in dealing with unforeseen challenges and ensuring mission success and crew safety.

5. Organizational Culture and Decision-Making:

The Challenger disaster exposed flaws in the organizational culture within NASA. It demonstrated that decision-making processes and management practices must be grounded in a strong safety culture, where individuals feel empowered to raise concerns without fear of reprisal and where safety considerations always take precedence.

6. Recognizing and Addressing Technical Flaws:

The technical failure of the O-rings in the solid rocket boosters served as a sobering reminder of the importance of thoroughly evaluating engineering designs and recognizing potential flaws. Understanding the limitations and vulnerabilities of critical components is essential to ensuring mission success and crew safety.

7. Learning from Failures:

The Challenger disaster demonstrated the necessity of learning from failures and mistakes. A culture that values continuous improvement and fosters a learning mindset is essential in advancing space exploration safely and effectively.

8. Respect for the Unpredictable Nature of Space:

Space exploration involves venturing into an unpredictable and harsh environment. The Challenger disaster reminded us of the need to respect and understand the challenges of spaceflight, continually updating our knowledge and technology to adapt to this unique environment.

9. Impact on Human Spaceflight Policy:

The Challenger disaster prompted reevaluation of human spaceflight policy. It led to changes in mission criteria, requirements for astronaut selection and training, and a greater focus on risk assessment in future missions.

10. Memorializing the Crew:

The loss of the Challenger crew members left an enduring impact on the space community. The tragedy has been memorialized in various ways, with educational programs, scholarships, and monuments dedicated to honoring the crew's memory and commitment to exploration.

The Challenger disaster was a somber reminder of the risks and challenges associated with space exploration. The lessons learned from this tragedy have significantly influenced space policy, safety practices, and the organizational culture within NASA. The crew's sacrifice continues to serve as a poignant reminder of the importance of safety, continuous improvement, and a steadfast commitment to the pursuit of knowledge and discovery in the vastness of space.

In the aftermath of the Space Shuttle Challenger disaster, NASA undertook significant safety reforms and implemented a series of changes to prevent future accidents and enhance the safety of human spaceflight. These reforms covered various aspects of mission planning, engineering, decision-making, and communication. Here are some of the key safety reforms that NASA put in place:

1. Redesign of Solid Rocket Boosters:

One of the most critical safety reforms was the redesign of the solid rocket boosters (SRBs). The O-rings, which had failed in the Challenger disaster, were modified to be more resilient to extreme temperatures. Additional safety measures were also introduced to enhance the overall performance and reliability of the SRBs.

2. Improved Risk Assessment and Decision-Making:

NASA implemented more robust risk assessment procedures and decision-making protocols. Mission managers were required to conduct more thorough evaluations of potential risks and address any safety concerns before proceeding with a launch. The emphasis on safety was prioritized over schedule considerations.

3. Enhanced Communication Channels:

To prevent communication breakdowns, NASA established clearer and more effective communication channels between engineers, contractors, and decision-makers. Open and transparent communication was encouraged, allowing concerns to be raised and addressed promptly.

4. Independent Oversight and Audits:

To ensure adherence to safety protocols and best practices, NASA introduced independent oversight and regular safety audits. This included external reviews and assessments by independent experts to provide an objective evaluation of safety procedures and recommendations for improvements.

5. Crew Training and Emergency Preparedness:

NASA enhanced crew training programs to include extensive emergency preparedness scenarios. Astronauts received comprehensive training on how to respond to potential emergencies during different phases of the mission, enabling them to act quickly and effectively in unforeseen situations.

6. Mission Management Team:

NASA established a Mission Management Team (MMT) responsible for overseeing the preparation and execution of space shuttle missions. This team consisted of top-level officials and technical experts, ensuring that decisions were made collaboratively, based on comprehensive data, and with a strong focus on safety.

7. Safety Review Panels:

The establishment of safety review panels allowed for thorough evaluations of mission-specific safety aspects and ensured that all potential risks were carefully considered. These panels helped in making informed decisions about the feasibility of each mission.

8. Reinforcement of Safety Culture:

NASA worked diligently to reinforce a strong safety culture within the agency. This included ongoing safety training for all personnel, fostering an environment where safety concerns were welcomed, and establishing a zero-tolerance policy for shortcuts that could compromise safety.

9. Advanced Technology and Data Analysis:

Advancements in technology and data analysis allowed NASA to gather more precise data and conduct more sophisticated risk assessments. This data-driven approach enabled the identification of potential safety issues before they became critical concerns.

10. Sharing Lessons Learned:

NASA actively shared the lessons learned from the Challenger disaster with other space agencies and organizations involved in human spaceflight. The aim was to promote a culture of continuous improvement and safety awareness throughout the global space community.

CHALLENGER: TRAGEDY AND TRIUMPH - UNRAVELING THE SPACE SHUTTLE CHALLENGER EXPLOSION

These safety reforms and improvements represented a fundamental shift in NASA's approach to human spaceflight. By learning from the Challenger disaster and continuously striving to enhance safety, NASA aimed to ensure the well-being of astronauts, the success of missions, and the advancement of space exploration. These reforms have contributed significantly to the agency's safety record and its ongoing commitment to exploring the cosmos while prioritizing the safety of its astronauts.

Chapter 8: The Human Factor

The Space Shuttle Challenger explosion had a profound and far-reaching psychological impact on the NASA community, the nation, and the global space community. The disaster, witnessed by millions of people through live television broadcast, created a collective sense of shock, grief, and loss that reverberated throughout society. The psychological effects were felt at multiple levels:

1. NASA Community:

Within the NASA community, the loss of the Challenger crew had a deeply emotional impact. Astronauts, engineers, and personnel who worked closely with the crew members experienced feelings of grief, sadness, and a sense of personal loss. The tragedy served as a stark reminder of the risks and sacrifices inherent in space exploration, challenging the agency's core beliefs and dedication to safety.

2. Families of the Crew:

The families of the Challenger crew members experienced an unimaginable and heart-wrenching loss. Their loved ones were national heroes, and the public outpouring of support added an additional layer of complexity to their grief. Coping with the loss of their family members while being in the spotlight of media attention was an overwhelming experience for the families.

3. National Trauma:

The Challenger explosion became a collective national trauma. The images of the shuttle disintegrating in the sky were etched into the memories of Americans, sparking feelings of sadness, disbelief, and a sense of shared loss. The tragedy led to a period of mourning and reflection across the nation.

4. Impact on Education and Aspirations:

The Challenger disaster deeply affected students and educators across the country. Christa McAuliffe, the "Teacher in Space," had captured the nation's imagination, and her tragic loss was a devastating blow to the educational community. The incident raised questions about the risks of space exploration and its potential impact on inspiring future generations to pursue careers in science and exploration.

5. Impact on Space Program:

The psychological impact of the Challenger disaster had a lasting effect on the space program. The suspension of shuttle missions for more than two years demonstrated the need for a thorough reevaluation of safety procedures and the importance of learning from the tragedy to prevent future accidents.

6. Fear and Uncertainty:

The Challenger explosion instilled fear and uncertainty about the safety of space travel in the minds of many people. Concerns about the risks of human spaceflight and doubts about the space program's ability to ensure astronaut safety were prevalent in public discourse.

7. Ethical and Moral Considerations:

The tragedy raised ethical and moral questions about the risks that astronauts willingly undertake in the pursuit of space exploration. Public discussions emerged about the balance between scientific progress, national pride, and the safety of human life.

8. Impact on the Space Community Globally:

The Challenger disaster resonated with the global space community, creating a sense of solidarity among space agencies worldwide. It highlighted the inherent dangers of space exploration and emphasized the importance of international cooperation in ensuring safety and the advancement of space science.

Despite the immense emotional toll, the Challenger disaster also spurred positive change. The commitment to safety and the thorough reforms implemented by NASA in response to the tragedy helped strengthen the agency's safety protocols and ensure the continued pursuit of space exploration with greater attention to the well-being of astronauts.

The Challenger explosion remains a poignant reminder of the courage and dedication of those who explore the cosmos and the importance of continually striving for safer, more reliable space travel. The memory of the lost crew members continues to serve as a symbol of humanity's relentless pursuit of knowledge and exploration while acknowledging the inherent risks and challenges of venturing into the unknown.

The Space Shuttle Challenger disaster had a profound and lasting impact on public perception of space exploration. The tragedy, witnessed by millions of people through live television broadcast, had several significant effects on how the general public viewed human spaceflight and the space program as a whole:

1. Heightened Awareness of Risks:

The Challenger disaster served as a stark reminder of the inherent risks and dangers associated with space exploration. For many, the incident shattered the illusion of space travel as routine or risk-free. It brought to the forefront the complexities and challenges of sending humans beyond Earth's atmosphere, leading to a more informed and cautious perception of space missions.

2. Skepticism and Safety Concerns:

Following the Challenger explosion, there was increased skepticism among the public regarding the safety of space missions. Many questioned the reliability of the shuttle program and the ability of NASA to ensure the safety of astronauts. The incident raised concerns about the decision-making process and the handling of potential risks.

3. Impact on Public Support:

The disaster had a significant impact on public support for the space program. While there was still admiration for the courage and dedication of astronauts, some segments of the public questioned the value and necessity of human spaceflight, especially in light of the risks involved.

4. Effects on Education and Inspiration:

The loss of Christa McAuliffe, the "Teacher in Space," had a profound impact on education and inspiration. Her presence on the Challenger mission had ignited excitement and hope for the future of space exploration and science education. Her tragic death led to concerns about the potential impact on inspiring the next generation of students to pursue careers in STEM fields.

5. National Mourning and Reflection:

The Challenger disaster led to a period of national mourning and reflection. It brought the entire nation together in collective grief and prompted discussions about the sacrifices made by astronauts and the risks they willingly undertake in the pursuit of knowledge and exploration.

6. Media Coverage and Public Memory:

The vivid and unforgettable images of the Challenger explosion were replayed repeatedly in the media, etching the tragedy into the public memory. The disaster's visibility in the media influenced public perception of space exploration and led to ongoing discussions about the risks and benefits of space missions.

7. Impact on Policy and Funding:

The public outcry and concerns following the Challenger disaster influenced space policy and funding decisions. The suspension of shuttle missions for over two years, along with the implementation of safety reforms, demonstrated the need for a more cautious and deliberate approach to human spaceflight.

8. Reinforcement of Safety Measures:

The disaster reinforced the importance of safety in the space program. NASA's subsequent focus on safety improvements and stringent risk assessment processes helped rebuild public confidence in the agency's commitment to astronaut safety.

The Challenger disaster had a complex and multi-faceted impact on public perception of space exploration. It led to heightened awareness of the risks and challenges of human spaceflight, skepticism about the reliability of space missions, and questions about the value and necessity of space exploration. However, it also prompted a national period of mourning and reflection, inspiring a more informed and cautious approach to space missions.

NASA's safety reforms and the agency's commitment to continuous improvement have since helped rebuild public confidence and reinforce the importance of safety in the pursuit of space exploration. The memory of the Challenger disaster continues to influence public discourse on the risks and rewards of venturing into the cosmos and serves as a somber reminder of the sacrifices made in the name of exploration and knowledge.

CHALLENGER: TRAGEDY AND TRIUMPH - UNRAVELING THE SPACE SHUTTLE CHALLENGER EXPLOSION

Chapter 9: Technological Advancements Post-Challenger

———

The Challenger disaster had a profound impact on spacecraft design and safety systems, prompting NASA and the space industry to implement significant technological advancements to enhance the safety and reliability of space missions. After the tragedy, numerous improvements were made to spacecraft design, engineering, materials, and safety protocols. Some of the key technological advancements include:

1. Solid Rocket Boosters (SRBs) Redesign:

The primary cause of the Challenger disaster was the failure of O-rings in the SRBs. In response, NASA redesigned the SRBs to include improved O-rings and additional safety features to prevent similar failures. The new design incorporated more robust materials and sealing mechanisms to ensure a reliable and secure seal during launch and ascent.

2. Enhanced Thermal Protection Systems (TPS):

The tragedy also highlighted the critical importance of thermal protection systems (TPS) on spacecraft. NASA developed and implemented improved TPS materials and techniques, such as advanced heat-resistant tiles and coatings, to safeguard the shuttle and other spacecraft from the intense heat generated during re-entry into Earth's atmosphere.

3. Abort and Escape Systems:

To improve crew safety during launch and ascent, spacecraft design incorporated advanced abort and escape systems. These systems allowed astronauts to evacuate the spacecraft in emergency situations, such as during a launch anomaly or malfunction, increasing the chances of survival for the crew.

4. Flight Control and Guidance Systems:

Advancements were made in spacecraft flight control and guidance systems to enhance stability, maneuverability, and precision during all phases of a mission. These improvements helped ensure smooth trajectories and safer flight profiles.

5. Advanced Avionics and Computers:

The Challenger disaster prompted upgrades to spacecraft avionics and computers. Advanced computer systems with redundancy and fault-tolerant features were integrated into spacecraft to improve system reliability and enhance the ability to handle contingencies.

6. Structural Integrity and Materials:

Post-Challenger, spacecraft engineers focused on improving structural integrity and materials used in spacecraft construction. Advanced composite materials and manufacturing techniques were employed to increase strength while reducing weight, enhancing overall vehicle performance and safety.

7. Risk Assessment and Management:

CHALLENGER: TRAGEDY AND TRIUMPH - UNRAVELING THE SPACE SHUTTLE CHALLENGER EXPLOSION

NASA implemented more comprehensive risk assessment and management procedures. Mission planners and managers evaluated potential risks more rigorously and made informed decisions based on thorough risk analyses.

8. Safety Training and Simulation:

Astronaut training programs were updated to include more extensive safety training and simulations of emergency scenarios. This ensured that the crew members were well-prepared to handle critical situations during the mission.

9. Independent Oversight and Safety Audits:

To ensure that safety protocols were followed rigorously, NASA introduced independent oversight and regular safety audits. External experts reviewed safety procedures and made recommendations for improvements.

10. International Collaboration:

The Challenger disaster also fostered international collaboration in space exploration. Sharing knowledge and experiences with other space agencies helped improve safety standards and best practices globally.

These technological advancements and safety improvements collectively transformed spacecraft design and operations, making space missions safer, more reliable, and more effective. The lessons learned from the Challenger disaster influenced the design of subsequent space vehicles, such as the Space Shuttle Endeavour, and informed the development of newer spacecraft like the Space Shuttle Atlantis, the Orion spacecraft, and the Commercial Crew vehicles.

The commitment to continuous improvement in spacecraft design and safety remains a core value of the space industry, ensuring that the lessons of the Challenger disaster have led to enduring advancements in human spaceflight.

CHALLENGER: TRAGEDY AND TRIUMPH - UNRAVELING THE SPACE SHUTTLE CHALLENGER EXPLOSION

Chapter 10: Return to Flight - STS-26

The journey of the Space Shuttle program's return to flight after the Challenger tragedy was a challenging and comprehensive process. The suspension of shuttle missions following the disaster in January 1986 initiated a period of deep introspection, safety reforms, and technological advancements to ensure the safe resumption of space shuttle missions. Here is an overview of the key steps and milestones on the road to the Space Shuttle program's return to flight:

1. Investigating the Disaster:

Immediately after the Challenger explosion, NASA formed the Rogers Commission, an independent committee tasked with investigating the cause of the tragedy. The commission's findings, released later in 1986, highlighted critical flaws in the shuttle program's safety protocols, decision-making processes, and organizational culture.

2. Implementing Safety Reforms:

NASA began implementing safety reforms based on the recommendations of the Rogers Commission. These reforms included redesigning the solid rocket boosters (SRBs) to enhance their safety, improving communication between engineers and decision-makers, and strengthening safety protocols and risk assessment procedures.

3. Improving Thermal Protection Systems (TPS):

One of the key lessons from the Challenger disaster was the need to enhance the shuttle's thermal protection system (TPS) to prevent heat-related accidents during re-entry. Advanced heat-resistant tiles and coatings were developed and integrated into the shuttle's heat shield to improve its resilience.

4. Testing and Validation:

Numerous tests and validation procedures were conducted to verify the effectiveness of the safety reforms and technological upgrades. The redesigned SRBs and improved TPS underwent extensive testing to ensure they could withstand the rigors of spaceflight.

5. Crew Safety Training:

Astronaut training programs were revamped to include more extensive safety training and simulations of emergency scenarios. Crew members underwent rigorous training to prepare for potential contingencies during the mission.

6. Close Examination of Shuttle Fleet:

Each shuttle in the fleet was meticulously inspected and overhauled to ensure its readiness for flight. Technicians and engineers conducted thorough checks of all systems and components, addressing any issues to ensure the shuttle's safety and performance.

7. STS-26: Return to Flight:

CHALLENGER: TRAGEDY AND TRIUMPH - UNRAVELING THE SPACE SHUTTLE CHALLENGER EXPLOSION

The first mission following the Challenger disaster was designated STS-26, and it marked the Space Shuttle program's return to flight. On September 29, 1988, Space Shuttle Discovery lifted off from Kennedy Space Center, carrying a crew of five astronauts. The successful completion of this mission was a significant milestone in restoring confidence in the space shuttle program.

8. Continued Vigilance:

Following STS-26, NASA remained vigilant in maintaining safety and addressing any potential concerns. The lessons learned from the Challenger disaster continued to inform mission planning and decision-making, with a renewed emphasis on safety and risk management.

9. Successive Missions:

Subsequent missions were conducted with a focus on safety and with the lessons of the Challenger disaster guiding every step. NASA continued to improve safety measures and technology, enhancing the shuttle's performance and reliability.

10. Legacy of Safety:

The safety reforms and technological advancements made in the aftermath of the Challenger disaster became a lasting legacy for the Space Shuttle program. NASA's commitment to safety and continuous improvement remained a cornerstone of its operations.

The journey of the Space Shuttle program's return to flight after the Challenger tragedy was marked by careful introspection, rigorous safety reforms, technological advancements, and a resolute commitment to honoring the memory of the lost crew members through safer and more successful space missions. The lessons learned from the disaster continue to inform space exploration practices, ensuring the safety of astronauts and the advancement of human spaceflight.

The STS-26 mission, designated as Space Shuttle Discovery's 7th flight, was a historic and highly anticipated event as it marked the Space Shuttle program's return to flight after the devastating Challenger explosion. Launched on September 29, 1988, STS-26 represented a critical milestone in restoring public confidence in the shuttle program and demonstrating NASA's commitment to safety and the advancement of human spaceflight.

Mission Crew:

The STS-26 crew consisted of five astronauts:

1. Frederick H. Hauck (Commander) - A veteran astronaut, Hauck had previously flown as the pilot of STS-7 and commander of STS-51-A.

2. Richard O. Covey (Pilot) - Covey was an experienced Air Force test pilot and made his first spaceflight on STS-26.

3. John M. Lounge (Mission Specialist) - Lounge, a career U.S. Air Force officer, was making his first spaceflight on this mission.

CHALLENGER: TRAGEDY AND TRIUMPH - UNRAVELING THE SPACE SHUTTLE CHALLENGER EXPLOSION

4. George D. Nelson (Mission Specialist) - Nelson, a scientist, had previously flown as a mission specialist on STS-41-C.

5. David C. Hilmers (Mission Specialist) - Hilmers, a Marine Corps officer, was making his second spaceflight, having previously flown on STS-51-J.

Mission Objectives:

The primary objective of STS-26 was to successfully deploy the Tracking and Data Relay Satellite-3 (TDRS-3) into orbit. TDRS-3 was part of a constellation of communication satellites that provided continuous tracking and data relay services for NASA's orbiting spacecraft.

Additionally, the mission aimed to conduct various scientific experiments and test new technologies to further the understanding of space and improve future shuttle missions.

Mission Highlights:

The launch of STS-26 from Kennedy Space Center in Florida was an emotionally charged event, as the nation and the world watched with bated breath. The successful liftoff signaled the resilience and determination of NASA and its commitment to safely resuming spaceflight.

During the mission, the crew members skillfully deployed TDRS-3, positioning it into its designated geostationary orbit. The satellite's successful deployment was a significant achievement, as it was crucial for enhancing communication and data relay capabilities for future space missions.

STS-26 also carried a Getaway Special (GAS) payload, which provided opportunities for various educational and scientific experiments. The crew conducted these experiments during their time in orbit, showcasing the versatility and scientific capabilities of the Space Shuttle.

One of the critical aspects of STS-26 was the thorough testing of various safety upgrades and technological advancements made to the shuttle's design after the Challenger disaster. The success of the mission validated the effectiveness of the safety reforms and boosted public confidence in the shuttle program.

The STS-26 mission was a pivotal moment in the history of the Space Shuttle program. The successful return to flight demonstrated NASA's resilience and dedication to space exploration in the face of adversity. It also underscored the agency's commitment to astronaut safety and continuous improvement in spacecraft design and operations. The legacy of STS-26 continues to inspire future generations of astronauts and space enthusiasts, reminding them of the courage and perseverance required to explore the cosmos.

CHALLENGER: TRAGEDY AND TRIUMPH - UNRAVELING THE SPACE SHUTTLE CHALLENGER EXPLOSION

Chapter 11: Challenger's Legacy in Education

———

The Space Shuttle Challenger's legacy in education is deeply intertwined with the tragic loss of Christa McAuliffe, the "Teacher in Space." Christa McAuliffe was a high school social studies teacher from Concord, New Hampshire, selected to be the first private citizen and educator to fly into space as part of the STS-51-L mission.

Her presence on the Challenger mission was intended to inspire students and educators worldwide, making space exploration and science more accessible and relatable to the general public. Despite the disaster, Christa McAuliffe's legacy in education and its influence on STEM (Science, Technology, Engineering, and Mathematics) programs has endured and left a lasting impact.

1. Inspiration and Aspiration:

The selection of a teacher to fly in space captured the imagination of the entire nation. Christa McAuliffe's journey from a high school classroom to space symbolized the power of education and the potential for anyone, regardless of their background, to contribute to the pursuit of knowledge and exploration. Her dream of going to space ignited the aspirations of countless students, inspiring them to pursue careers in STEM fields and reach for their own dreams.

2. STEM Advocacy:

The tragedy brought renewed focus on the importance of STEM education and advocacy. In the wake of the Challenger disaster, numerous initiatives were launched to promote science education, particularly in underserved communities. The aim was to foster a new generation of scientists, engineers, and explorers who could continue the spirit of discovery that Christa McAuliffe embodied.

3. Educational Outreach Programs:

In honor of Christa McAuliffe's memory, NASA and other organizations established educational outreach programs to engage students and educators in science and space exploration. These programs offered a range of educational resources, including lesson plans, workshops, and hands-on activities, to inspire students' interest in space and science.

4. Challenger Centers:

The Challenger Center for Space Science Education was established as a living tribute to the Challenger crew, with a focus on space education and hands-on learning experiences. The centers offer simulated space missions and interactive STEM activities, providing students with an immersive and engaging space exploration experience.

5. McAuliffe-Shepard Discovery Center:

The McAuliffe-Shepard Discovery Center in Concord, New Hampshire, was established to celebrate the legacies of both Christa McAuliffe and astronaut Alan Shepard. The center features interactive exhibits and educational programs that promote space science, aviation, and STEM education.

6. Scholarships and Grants:

Various scholarships and grants were created in memory of the Challenger crew members, including Christa McAuliffe. These scholarships aim to support students pursuing STEM-related fields and honor the crew's commitment to education and exploration.

7. Impact on STEM Teachers:

The tragedy also highlighted the vital role of educators in inspiring future scientists and engineers. Teachers, like Christa McAuliffe, are essential in nurturing curiosity and encouraging students to pursue STEM careers. The Challenger disaster reinforced the importance of supporting STEM teachers and providing them with resources and opportunities for professional development.

The Challenger's legacy in education is a powerful reminder of the enduring impact of science, exploration, and the human spirit. While the tragedy claimed the lives of brave astronauts, including Christa McAuliffe, it also sparked a wave of renewed enthusiasm for STEM education and advocacy.

The Challenger disaster reinforced the importance of education in space exploration, inspiring generations of students to pursue careers in science and engineering and fostering a deep appreciation for the wonders of the universe. The legacy of Christa McAuliffe and the Challenger crew lives on in the hearts and minds of those who continue to explore the frontiers of knowledge and strive to make the world a better place through the pursuit of science and discovery.

The Teacher in Space Project, initiated by NASA in the 1980s, had a profound and lasting impact on education, space exploration, and the public's perception of science and technology. Although the project tragically ended with the loss of Christa McAuliffe and the Challenger crew, its legacy continues to inspire and shape educational initiatives and the promotion of STEM (Science, Technology, Engineering, and Mathematics) fields. Here are some key impacts of the Teacher in Space Project:

1. Inspiring Students and Educators:

The selection of a teacher to fly into space captured the imagination of students and educators worldwide. Christa McAuliffe's dream of being the first teacher in space ignited the aspirations of countless students to pursue careers in STEM fields. The project showed that science and space exploration were not distant endeavors but within reach for anyone with dedication and passion.

2. Elevating the Importance of STEM Education:

The Teacher in Space Project brought renewed attention to the significance of STEM education in preparing the next generation for future challenges and opportunities. The project underscored the role of teachers as critical influencers in sparking curiosity and nurturing scientific interest in young minds.

3. Fostering Public Interest in Space Exploration:

The involvement of a civilian teacher in a space mission generated widespread media coverage and public interest. The project helped bridge the gap between space exploration and the general public, making space science and technology more accessible and relevant to everyday lives.

4. Raising Awareness of Space Missions:

The Teacher in Space Project brought attention to the Space Shuttle program and the significance of its missions. It highlighted the practical applications of space science and the potential for space exploration to benefit humanity.

5. Establishing Educational Outreach Programs:

In the aftermath of the Challenger disaster, several educational outreach programs were established in memory of Christa McAuliffe and the Teacher in Space Project. These programs aimed to honor the crew's commitment to education and inspire the next generation of scientists and engineers.

6. Challenger Centers for Space Science Education:

The Challenger Centers, created as part of the Teacher in Space Project's legacy, are interactive learning centers that offer students and educators simulated space missions and hands-on STEM activities. These centers continue to engage students in space exploration and inspire a passion for science.

7. Memorializing Christa McAuliffe's Legacy:

The memory of Christa McAuliffe and her commitment to education and space exploration continues to be honored through various scholarships, awards, and memorials. Her legacy serves as a reminder of the importance of educators in shaping the future of scientific exploration.

8. Impact on Space Policy and Education Initiatives:

The Teacher in Space Project influenced space policy and education initiatives, sparking discussions about the role of civilians in space missions and the value of education in the pursuit of space exploration. It inspired a greater emphasis on STEM education and the inclusion of educational components in space missions.

The Teacher in Space Project left a profound impact on education, space exploration, and public perception of science and technology. Despite the tragedy that cut short the project, its legacy has continued to inspire students and educators, promoting the importance of STEM education, fostering interest in space exploration, and encouraging a sense of wonder and curiosity about the universe. The project's emphasis on the intersection of education and space exploration continues to shape educational initiatives and underscore the power of teachers in igniting the flame of scientific inquiry in future generations.

CHALLENGER: TRAGEDY AND TRIUMPH - UNRAVELING THE SPACE SHUTTLE CHALLENGER EXPLOSION

Chapter 12: Space Shuttle Program and Beyond

The Challenger explosion had a significant and wide-ranging impact on the Space Shuttle program, affecting various aspects of its operations, safety protocols, public perception, and overall direction. The disaster, which occurred on January 28, 1986, during the STS-51-L mission, represented a major turning point in the history of human spaceflight and left a lasting legacy on the space program. Here are some key areas where the Challenger explosion had an overall impact on the Space Shuttle program:

1. Safety Reforms and Technological Advancements:

The Challenger disaster exposed critical flaws in the Space Shuttle program's safety protocols and design. In response, NASA undertook comprehensive safety reforms and implemented technological advancements to prevent future accidents. These reforms included redesigning the solid rocket boosters (SRBs), improving the shuttle's thermal protection system (TPS), enhancing safety procedures, and implementing rigorous risk assessments.

2. Suspension of Shuttle Missions:

Following the Challenger explosion, NASA suspended shuttle missions for over two years. This hiatus allowed the agency to thoroughly investigate the accident, address safety concerns, and implement necessary changes to ensure the safe return to flight.

3. Shift in Public Perception:

The Challenger disaster had a profound impact on public perception of space exploration and the Space Shuttle program. It raised questions about the risks involved in human spaceflight and led to increased scrutiny of NASA's decision-making and safety practices. Public confidence in the shuttle program was shaken, and skepticism regarding the safety of space missions became more prevalent.

4. Impact on Crewed Spaceflight Policy:

The disaster prompted a reevaluation of crewed spaceflight policy. It led to changes in mission criteria, requirements for astronaut selection and training, and a greater emphasis on risk assessment and management in future missions.

5. Organizational Culture and Decision-Making:

The Challenger explosion exposed flaws in NASA's organizational culture and decision-making processes. The focus on meeting schedules and the pressure to launch contributed to a disregard for safety concerns. The tragedy highlighted the importance of fostering a safety-first culture and creating an environment where concerns can be raised without fear of reprisal.

6. Legacy of the Challenger Crew:

CHALLENGER: TRAGEDY AND TRIUMPH - UNRAVELING THE SPACE SHUTTLE CHALLENGER EXPLOSION

The legacy of the Challenger crew members, who made the ultimate sacrifice in the pursuit of exploration, continues to be a source of inspiration and remembrance. Their names and their commitment to space exploration are memorialized in various ways, ensuring that their memory endures.

7. Shuttle Program's Return to Flight:

The process of returning the Space Shuttle program to flight was arduous, involving extensive testing, validation, and safety checks. The successful return to flight of Space Shuttle Discovery on the STS-26 mission in September 1988 marked a significant milestone in rebuilding public confidence in the shuttle program.

8. Lessons for Future Space Programs:

The Challenger disaster provided valuable lessons for future space programs and missions. The importance of safety, the need for thorough risk assessment, and the significance of a strong safety culture were all key takeaways from the tragedy. These lessons have informed subsequent space exploration endeavors and influenced the approach to crewed spaceflight worldwide.

The Challenger explosion had a profound and lasting impact on the Space Shuttle program. The tragedy led to safety reforms, technological advancements, and changes in space policy. It highlighted the challenges and risks associated with human spaceflight and reshaped public perception of space exploration. The memory of the Challenger crew members and their sacrifice continues to serve as a poignant reminder of the importance of safety, continuous improvement, and unwavering dedication to the pursuit of knowledge and discovery in the vastness of space.

After the Challenger accident, NASA underwent significant changes in its focus and approach to future missions. The tragedy served as a wake-up call, prompting a reevaluation of safety practices, mission priorities, and organizational culture. Several key shifts in NASA's focus and future missions emerged in response to the Challenger disaster:

1. Enhanced Safety and Risk Management:

The Challenger explosion underscored the critical importance of safety in human spaceflight. NASA became much more risk-conscious and implemented robust safety protocols. Safety considerations took precedence over schedule pressures, and thorough risk assessments became integral to mission planning and decision-making.

2. Retirement of the Space Shuttle Program:

While the Space Shuttle program resumed missions after the Challenger disaster, it eventually came to an end decades later. In the wake of the accident, NASA recognized the limitations and vulnerabilities of the shuttle's design. The focus shifted toward developing a safer and more advanced space transportation system for future missions.

3. Development of the International Space Station (ISS):

The Challenger disaster accelerated the need for international collaboration in space exploration. NASA joined forces with international partners, including Russia, Europe, Japan, and Canada, to develop the International Space Station (ISS). The ISS became a symbol of cooperation in space and provided a platform for scientific research, technology development, and long-duration human spaceflight.

4. Unmanned Space Missions:

NASA shifted some of its focus toward unmanned space missions, particularly robotic exploration of other planets and celestial bodies. Unmanned missions offered valuable scientific data and paved the way for future human exploration beyond Earth.

5. Exploration of Mars:

The Challenger disaster contributed to a rekindled interest in exploring Mars. NASA redirected its vision toward crewed missions to the Red Planet, with robotic missions serving as precursors to human exploration.

6. Commercial Spaceflight:

The Challenger disaster fueled discussions about the role of private companies in space exploration. Over time, NASA shifted its approach to foster commercial spaceflight capabilities. Collaborating with private companies for cargo and crew transportation to the ISS became a key component of NASA's spaceflight strategy.

7. Focus on Scientific Research:

With the expansion of unmanned space missions, NASA intensified its focus on scientific research. Missions were tailored to study Earth, the solar system, and the universe, deepening our understanding of astrophysics, planetary science, climate change, and more.

8. Outreach and Education:

The Challenger tragedy reinforced the importance of public outreach and education in space exploration. NASA redoubled its efforts to engage with the public, inspire the next generation of scientists and engineers, and foster a sense of wonder about the cosmos.

9. Artemis Program:

In recent years, NASA has embarked on the Artemis program, aiming to return humans to the Moon and pave the way for future crewed missions to Mars. The program embodies the lessons learned from past accidents and emphasizes safety, international collaboration, and the exploration of new frontiers.

The Challenger accident triggered significant shifts in NASA's focus and future missions. The disaster led to a renewed emphasis on safety, international cooperation, and scientific exploration. It encouraged NASA to explore new possibilities in space while also fostering a greater sense of responsibility and vigilance in ensuring the success and safety of human spaceflight. The Challenger tragedy's legacy continues to influence NASA's mission planning and execution, guiding the agency toward a future of exploration and discovery in the cosmos.

OLIVER LANCASTER

Chapter 13: Remembering Challenger - Memorials and Tributes

———

The Challenger crew members, who tragically lost their lives on January 28, 1986, during the STS-51-L mission, have been honored and memorialized in various ways to pay tribute to their bravery, dedication, and sacrifice. Over the years, numerous memorials and tributes have been dedicated to the Challenger crew, ensuring that their memory endures. Here are some notable examples:

1. Space Mirror Memorial (Kennedy Space Center, Florida):

The Space Mirror Memorial, also known as the Space Mirror Memorial Wall, is located at the Kennedy Space Center Visitor Complex in Florida. It features the names of all 24 astronauts who lost their lives in the line of duty during space missions, including the seven Challenger crew members. The mirror's polished granite surface reflects the Space Shuttle Challenger Monument and serves as a poignant reminder of the risks and sacrifices of space exploration.

2. Challenger Center for Space Science Education:

The Challenger Center for Space Science Education was established as a living tribute to the Challenger crew. This nonprofit organization was founded by the families of the astronauts to continue the crew's educational mission. The Challenger Centers, located in various regions, offer simulated space missions and interactive STEM experiences for students, providing hands-on learning opportunities and inspiring the next generation of explorers.

3. Challenger Learning Centers:

Challenger Learning Centers are part of the Challenger Center for Space Science Education and are dedicated to engaging students and educators in space science and exploration. These centers offer simulated space missions and workshops that promote teamwork, problem-solving, and critical thinking, embodying the spirit of the Challenger crew's dedication to education.

4. Christa McAuliffe Planetarium (Concord, New Hampshire):

The Christa McAuliffe Planetarium, located in Concord, New Hampshire, is named in honor of Christa McAuliffe, the "Teacher in Space." The planetarium provides educational programs and astronomy-related events, fostering an interest in science and space exploration among students and the public.

5. McAuliffe-Shepard Discovery Center (Concord, New Hampshire):

The McAuliffe-Shepard Discovery Center, also in Concord, New Hampshire, is dedicated to celebrating the legacies of Christa McAuliffe and astronaut Alan Shepard. The center features interactive exhibits on space science, aviation, and STEM education.

6. National Challenger Center (Arlington, Virginia):

CHALLENGER: TRAGEDY AND TRIUMPH - UNRAVELING THE SPACE SHUTTLE CHALLENGER EXPLOSION

The National Challenger Center, situated in Arlington, Virginia, is an organization that offers educational programs and resources to inspire students and educators in the fields of science, technology, engineering, and mathematics. It honors the Challenger crew's commitment to education and exploration.

7. Christa Corrigan McAuliffe Elementary School (Houston, Texas):

In Houston, Texas, an elementary school was named in memory of Christa McAuliffe. The school's name serves as a reminder of the importance of education and the enduring impact of the Teacher in Space Project.

8. Challenger Seven Memorial Park (Orange, Texas):

The Challenger Seven Memorial Park in Orange, Texas, is dedicated to the memory of the Challenger crew. The park features a bronze memorial sculpture and plaques with the crew members' names, serving as a tribute to their lives and legacy.

These memorials and tributes stand as lasting reminders of the Challenger crew's contributions and sacrifice, inspiring generations to embrace the spirit of exploration, education, and discovery in the quest to expand humanity's understanding of the universe.

The memory of the Challenger crew lives on in space exploration through several enduring aspects that continue to inspire and influence the pursuit of knowledge and discovery beyond Earth:

1. Commitment to Education and Outreach:

The Challenger crew's dedication to education and outreach has left a lasting impact on space exploration. Their legacy inspired the creation of educational programs, learning centers, and initiatives aimed at engaging students and educators in science, technology, engineering, and mathematics (STEM). The Challenger Centers and similar educational organizations continue to foster a passion for space science and exploration among future generations.

2. Emphasis on Safety and Continuous Improvement:

The Challenger disaster served as a poignant reminder of the importance of safety in space exploration. Their sacrifice reinforced NASA's commitment to enhancing safety protocols, risk assessments, and technology. The lessons learned from the tragedy continue to shape the culture of space agencies worldwide, ensuring that safety remains paramount in all space missions.

3. Exploration of New Frontiers:

The Challenger crew's willingness to venture into the unknown and explore new frontiers echoes in current and future space exploration endeavors. Their bravery inspires astronauts and scientists to push the boundaries of human knowledge, exploring the Moon, Mars, and beyond.

4. Human Spaceflight Legacy:

The Challenger crew's legacy resonates in the accomplishments of subsequent human spaceflight missions. Their memory is honored in every astronaut who embarks on a journey to space, carrying their spirit of adventure and commitment to exploration.

5. Space Policy and Decision-Making:

The Challenger disaster prompted a reevaluation of space policy and decision-making processes. The tragic event influenced the criteria for astronaut selection, the prioritization of safety, and the consideration of risks in future space missions.

6. Public Engagement and Inspiration:

The Challenger crew's story captured the hearts of people around the world. Their bravery and dedication to education continue to inspire millions of individuals to embrace the wonders of space and pursue careers in science, engineering, and space-related fields.

7. Space Exploration Memorials:

Various memorials, such as the Space Mirror Memorial and the Challenger Seven Memorial Park, stand as tangible reminders of the Challenger crew's sacrifice and the risks inherent in space exploration. These memorials serve as symbols of remembrance and inspire reflection on the human quest for knowledge and understanding.

8. Ongoing Missions and Discoveries:

The Challenger crew's memory is honored each time a new space mission sets out to explore the cosmos. Every new discovery, whether through telescopes, robotic missions, or crewed expeditions, is a testament to the enduring spirit of exploration embodied by the Challenger crew.

The memory of the Challenger crew lives on in space exploration through their enduring legacy of education, safety, and exploration. Their sacrifice continues to shape space policy, inspire future generations, and remind humanity of the importance of pushing the boundaries of knowledge and understanding. Their memory serves as a guiding light for all who venture into the cosmos, carrying the spirit of exploration and the pursuit of new horizons.

CHALLENGER: TRAGEDY AND TRIUMPH - UNRAVELING THE SPACE SHUTTLE CHALLENGER EXPLOSION

Chapter 14: The Shuttle Program's End

The final years of the Space Shuttle program were marked by a series of significant milestones, missions, and bittersweet farewells. As NASA prepared to retire the Space Shuttle fleet, the focus shifted towards completing the remaining missions and transitioning to the next era of human spaceflight. Here is a chronicle of the final years of the Space Shuttle program:

2010:

- STS-130 (Endeavour): Launched on February 8, 2010, this mission delivered the Tranquility module and the Cupola to the International Space Station (ISS). The Cupola provided astronauts with a panoramic view of Earth and space.

- STS-131 (Discovery): Launched on April 5, 2010, this mission delivered essential supplies and equipment to the ISS. The crew conducted several spacewalks to perform maintenance and install new components.

- STS-132 (Atlantis): Launched on May 14, 2010, this mission delivered the Russian Mini-Research Module-1 to the ISS. The module provided additional storage and docking space.

- STS-133 (Discovery): Launched on February 24, 2011, this mission delivered the Permanent Multipurpose Module and the humanoid robot Robonaut 2 to the ISS.

- STS-134 (Endeavour): Launched on May 16, 2011, this mission delivered the Alpha Magnetic Spectrometer (AMS) to the ISS. The AMS was designed to study cosmic rays and search for antimatter and dark matter.

- STS-135 (Atlantis): Launched on July 8, 2011, this was the final Space Shuttle mission. The crew delivered supplies and equipment to the ISS, marking the end of an era in human spaceflight.

July 21, 2011:

- Atlantis landed at Kennedy Space Center, marking the conclusion of the Space Shuttle program after 30 years of operation. With this final landing, the Space Shuttle fleet was retired.

Post-Shuttle Program:

- After the retirement of the Space Shuttle program, NASA shifted its focus to developing the next generation of spacecraft, including the Orion spacecraft for deep space exploration and partnering with commercial space companies for crewed missions to the ISS.

- The end of the Space Shuttle program led to a gap in U.S. human spaceflight capabilities, as NASA relied on Russian Soyuz spacecraft to transport astronauts to and from the ISS until the development of new American crewed spacecraft.

CHALLENGER: TRAGEDY AND TRIUMPH - UNRAVELING THE SPACE SHUTTLE CHALLENGER EXPLOSION

- The legacy of the Space Shuttle program continues to influence space exploration, as it facilitated the construction of the ISS, conducted groundbreaking scientific research, and served as a workhorse in space missions for three decades.

Overall, the final years of the Space Shuttle program were a time of celebration and reflection on its historic achievements, as well as a period of transition and preparation for the future of human spaceflight. The program's legacy continues to inspire space agencies and private companies to push the boundaries of exploration and pave the way for future missions beyond Earth's orbit.

The decision to retire the Space Shuttle fleet was driven by a combination of factors, including safety concerns, high operational costs, and the need to shift focus towards the development of new space exploration capabilities. The implications of retiring the Space Shuttle fleet were both significant and multi-faceted, impacting various aspects of space exploration and human spaceflight. Here are some key reasons behind the decision to retire the shuttle fleet and its implications:

1. Safety Concerns:

The Space Shuttle program had experienced two catastrophic accidents, the Challenger explosion in 1986 and the Columbia disaster in 2003, which resulted in the loss of 14 astronauts. These tragedies highlighted the inherent risks of human spaceflight and raised concerns about the safety of the aging shuttle fleet. NASA recognized the need to prioritize crew safety and minimize risks in future missions.

2. Aging Fleet and High Operational Costs:

By the 2010s, the Space Shuttle fleet was nearing the end of its operational lifespan. The shuttles were expensive to maintain, refurbish, and launch. The cost of each shuttle mission was significantly higher compared to other space launch systems, such as unmanned rockets. NASA's budget constraints led to the consideration of more cost-effective alternatives for transporting crew and cargo to space.

3. Need for Next-Generation Spacecraft:

NASA sought to transition from the Shuttle program to next-generation spacecraft that could facilitate future exploration beyond low Earth orbit. The agency aimed to develop spacecraft capable of crewed missions to destinations like the Moon, Mars, and beyond. Shifting away from the Shuttle allowed NASA to focus its resources and efforts on developing these new space exploration capabilities.

4. Dependence on International Partners and Commercial Spaceflight:

With the retirement of the Shuttle fleet, NASA temporarily relied on Russian Soyuz spacecraft to transport astronauts to and from the International Space Station (ISS). However, the long-term goal was to develop American crewed spacecraft through partnerships with commercial spaceflight companies.

The shift towards commercial crew capabilities aimed to increase access to space and foster a sustainable space economy.

5. Focus on Deep Space Exploration:

Retiring the Space Shuttle fleet allowed NASA to redirect its focus towards deep space exploration missions. The agency's priority shifted towards developing the Orion spacecraft and the Space Launch System (SLS), designed for crewed missions beyond low Earth orbit, including lunar and Martian exploration.

Implications for Space Exploration:

1. Transition to New Spacecraft: The retirement of the Shuttle fleet necessitated the development of new spacecraft and launch systems for crewed missions. The focus shifted towards the Orion spacecraft and SLS, which would be essential for future human space exploration missions.

2. Enhanced International Collaboration: The reliance on Russian Soyuz spacecraft during the post-Shuttle era underscored the importance of international collaboration in space exploration. NASA's partnership with other space agencies, including Russia, European Space Agency, Japan, and Canada, strengthened the global effort in human spaceflight and scientific research on the ISS.

3. Commercial Crew Capabilities: The decision to retire the Shuttle fleet encouraged the growth of commercial spaceflight capabilities. NASA's collaborations with private companies, such as SpaceX and Boeing, led to the development of crewed

spacecraft like the Crew Dragon and CST-100 Starliner, enabling the United States to regain the capability of launching astronauts from American soil.

4. Emphasis on Deep Space Exploration: With the Shuttle program concluded, NASA's emphasis shifted to deep space exploration missions, including crewed missions to the Moon and Mars. The development of new spacecraft and launch systems supported the agency's vision for human exploration beyond Earth orbit.

5. Inspiration and Innovation: The retirement of the Shuttle fleet marked the end of an era and triggered a new phase of space exploration. The transition towards new spacecraft, international collaboration, and commercial space endeavors sparked inspiration, innovation, and excitement in the space community and the general public.

The decision to retire the Space Shuttle fleet was a pivotal moment in the history of human spaceflight. It reflected a strategic shift in NASA's approach to space exploration, prioritizing safety, international collaboration, and the development of next-generation spacecraft. The implications of retiring the Shuttle fleet continue to shape space exploration efforts, encouraging advancements in technology, fostering international cooperation, and setting the stage for future crewed missions to explore the cosmos and extend humanity's presence beyond Earth.

CHALLENGER: TRAGEDY AND TRIUMPH - UNRAVELING THE SPACE SHUTTLE CHALLENGER EXPLOSION

Chapter 15: Challenger's Lasting Impact on Safety Culture

———

The Challenger disaster, which occurred on January 28, 1986, had a profound and lasting impact on safety culture within NASA and the aerospace industry as a whole. The tragedy exposed critical flaws in decision-making, organizational culture, and safety protocols, prompting a comprehensive reevaluation of practices to prevent future accidents. Here are key ways in which the Challenger disaster transformed safety culture within NASA and the aerospace industry:

1. Priority of Safety over Schedule:

Before the Challenger disaster, there was a prevailing culture at NASA that prioritized meeting schedules and mission timelines, sometimes at the expense of safety concerns. The accident underscored the need to reverse this approach and established safety as the top priority in all aspects of spaceflight. NASA and the aerospace industry recognized that missions should proceed only when the risks are minimized and acceptable.

2. Redefining Risk Assessment and Management:

The Challenger disaster revealed a failure to adequately assess and communicate the risks associated with the Space Shuttle's solid rocket boosters (SRBs) in cold weather conditions. The accident led to a more rigorous approach to risk assessment and management, involving comprehensive analysis, testing, and contingency planning to address potential hazards.

3. Improved Communication and Decision-Making:

The tragedy exposed flaws in communication between engineering teams and management. Engineers who raised safety concerns felt their warnings were not taken seriously. In response, NASA and the aerospace industry initiated efforts to foster open communication channels, ensuring that safety-critical information is shared and considered in decision-making processes.

4. Focus on Engineering and Technical Excellence:

The Challenger disaster highlighted the importance of technical expertise and engineering excellence in spaceflight. The aerospace industry recognized that maintaining the highest standards of engineering, design, and testing was essential to ensuring the safety and success of missions.

5. Safety Culture Transformation:

The accident prompted a fundamental transformation in safety culture within NASA and the aerospace industry. Organizations adopted a proactive approach to safety, encouraging employees at all levels to identify and report potential risks or safety issues. Safety became an integral part of the organization's values, with a commitment to continuous improvement and learning from past mistakes.

6. Training and Professional Development:

CHALLENGER: TRAGEDY AND TRIUMPH - UNRAVELING THE SPACE SHUTTLE CHALLENGER EXPLOSION

Following the Challenger disaster, NASA and the aerospace industry invested more resources in training and professional development for employees. Astronauts, engineers, and support staff received enhanced safety training, technical instruction, and simulations to better prepare for potential emergency scenarios.

7. Independent Oversight and Review:

To enhance safety and prevent potential conflicts of interest, the aerospace industry established independent oversight and review boards. These boards provided impartial assessments of mission readiness, safety concerns, and technical challenges, offering an additional layer of scrutiny and accountability.

8. Impact on Future Accidents:

The lessons learned from the Challenger disaster had implications beyond that specific tragedy. The safety culture transformation spurred by the accident helped shape responses and decision-making in subsequent accidents, such as the Space Shuttle Columbia disaster in 2003. The focus on safety and risk management became an integral part of investigating and addressing future incidents.

The Challenger disaster was a pivotal moment in the history of space exploration, prompting a profound transformation in safety culture within NASA and the aerospace industry. The tragedy served as a powerful reminder of the risks and challenges inherent in human spaceflight and emphasized the critical importance of prioritizing safety in all space missions. The lessons learned from the Challenger disaster continue to inform safety practices and decision-making in space exploration endeavors, contributing to a safer and more resilient aerospace industry.

CHALLENGER: TRAGEDY AND TRIUMPH - UNRAVELING THE SPACE SHUTTLE CHALLENGER EXPLOSION

Chapter 16: Lessons for Future Space Exploration

———

The long-term impact of the Challenger explosion on space exploration missions beyond the Shuttle program has been far-reaching and has significantly influenced the trajectory of human spaceflight and exploration. The tragedy, which occurred on January 28, 1986, prompted a comprehensive reevaluation of safety protocols, mission planning, and the development of next-generation spacecraft. Here are key ways in which the Challenger explosion shaped space exploration missions beyond the Shuttle program:

1. Emphasis on Safety and Risk Management:

The Challenger disaster underscored the critical importance of safety in space exploration. It initiated a paradigm shift that prioritized crew safety and risk management in all aspects of human spaceflight. Space agencies and private companies engaged in space exploration missions beyond the Shuttle program have since adopted a proactive approach to safety, conducting thorough risk assessments, and developing contingency plans to minimize potential hazards.

2. Development of New Spacecraft and Technologies:

In the aftermath of the Challenger explosion, space agencies focused on developing new spacecraft and technologies for future exploration missions. The need for safe, reliable, and advanced spacecraft became a priority. This led to the development of spacecraft like the Orion crew capsule and the Space Launch System (SLS) in the United States, both designed for deep space exploration beyond Earth orbit.

3. International Collaboration and Partnerships:

The Challenger disaster emphasized the significance of international collaboration in space exploration. Space agencies worldwide recognized the benefits of pooling resources, expertise, and technology to advance exploration missions beyond the Shuttle program. The International Space Station (ISS) is a prime example of successful international collaboration, with various nations contributing modules and crew members to conduct research and experiments in microgravity.

4. Focus on Deep Space Exploration:

The Challenger explosion heightened the focus on deep space exploration missions, such as crewed missions to the Moon, Mars, and beyond. The tragic event served as a catalyst for reimagining the future of space exploration and pursuing ambitious missions beyond the confines of low Earth orbit.

5. Commercial Space Exploration:

The Challenger disaster influenced the development of commercial space exploration. With the Shuttle program's retirement, space agencies sought to partner with private companies for crewed and uncrewed missions. Commercial spaceflight companies, like SpaceX and Boeing, have played a crucial role in developing crewed spacecraft, resupplying the ISS, and driving innovation in space exploration.

6. Space Policy and Funding:

The Challenger explosion had implications on space policy and funding. It led to discussions on the balance between ambitious exploration goals and the need for safety and risk mitigation. Government agencies and lawmakers reconsidered the level of funding required for space exploration missions, particularly for long-duration missions to distant celestial bodies.

7. Legacy of Crew Memorials:

The memory of the Challenger crew members continues to resonate in space exploration missions beyond the Shuttle program. Crew memorials, educational programs, and initiatives continue to honor their sacrifice and dedication to space exploration, serving as a reminder of the human spirit of discovery.

The Challenger explosion had a profound and lasting impact on space exploration missions beyond the Shuttle program. It shaped the focus on safety, the development of new spacecraft and technologies, and the emphasis on international collaboration in space exploration. The tragedy served as a catalyst for redefining the future of human spaceflight, with a renewed focus on deep space exploration and the emergence of commercial space exploration. The Challenger crew's memory lives on in the continued quest to explore the cosmos, expand humanity's horizons, and push the boundaries of knowledge in the vastness of space.

The lessons learned from the Challenger disaster continue to have a profound and lasting impact on shaping future space missions. The tragedy served as a powerful catalyst for reevaluating safety protocols, risk management practices, and decision-making processes in space exploration. As space agencies and private companies plan and execute missions, the following are some key ways the lessons from Challenger influence future endeavors:

1. Safety as the Top Priority:

The foremost lesson from the Challenger disaster was the importance of prioritizing safety above all else. Space agencies and private companies have embraced a safety-first approach, implementing stringent safety protocols, thorough risk assessments, and continuous monitoring of potential hazards. Every mission is subject to rigorous safety checks to ensure the well-being of crew members and mission success.

2. Risk Mitigation and Contingency Planning:

The Challenger explosion highlighted the necessity of anticipating and mitigating risks in space missions. Future missions now involve comprehensive risk assessments and the development of contingency plans to address potential issues. These plans ensure that astronauts and spacecraft are prepared for various scenarios, enhancing mission resilience and adaptability.

3. Engineering Excellence and Technical Expertise:

The tragedy emphasized the critical role of engineering excellence and technical expertise in space exploration. Future missions place a strong emphasis on maintaining the highest standards of engineering design, testing, and manufacturing. The use of advanced technologies and innovative solutions enhances the reliability and performance of spacecraft and equipment.

4. Open Communication and Transparency:

The Challenger disaster revealed the importance of open communication and transparency in mission planning and decision-making. A culture of open dialogue encourages engineers and professionals to voice safety concerns without fear of reprisal. Lessons from Challenger have prompted a more collaborative and transparent environment in which critical information is shared and considered at all levels of the organization.

5. International Collaboration:

The disaster highlighted the benefits of international collaboration in space exploration. Space agencies from different countries work together on complex missions, leveraging their collective resources, expertise, and capabilities. International partnerships ensure the sharing of knowledge and risks, leading to more robust and successful missions.

6. Emphasis on Continuous Learning:

The lessons from Challenger have instilled a commitment to continuous learning and improvement in space missions. Space agencies and companies conduct post-mission reviews, analyze data, and share findings to learn from past experiences and optimize future missions. These reviews lead to refinements in processes and safety measures, driving progress in space exploration.

7. Spacecraft Design and Redundancy:

Future missions have incorporated the lessons learned from Challenger into spacecraft design. Redundancy is built into critical systems to ensure that the failure of a single component does not jeopardize the entire mission. Robust designs, redundant safety systems, and fail-safe measures contribute to mission success and crew safety.

8. Public Perception and Outreach:

The Challenger disaster had a profound impact on public perception of space exploration. The lessons learned have led to increased efforts in public outreach and communication. Space agencies and companies actively engage with the public to foster understanding, support, and interest in space missions.

The lessons learned from the Challenger disaster continue to shape future space missions in profound ways. Safety remains the paramount consideration, influencing all aspects of mission planning and execution. The emphasis on risk mitigation, engineering excellence, open communication, and international collaboration has resulted in safer and more successful space exploration endeavors.

As we push the boundaries of human knowledge and journey deeper into the cosmos, the lessons from Challenger serve as a guiding beacon, ensuring that the pursuit of space exploration remains steadfast in its commitment to safety, innovation, and the betterment of humanity.

OLIVER LANCASTER

Chapter 17: Historical and Cultural Impact

The Challenger explosion became a significant event in American history due to its tragic and highly public nature. The disaster unfolded on live television, witnessed by millions of Americans and people around the world. It was a pivotal moment that shook the nation, leading to profound cultural impacts and shaping the trajectory of space exploration and public perception. Here are key aspects of how the Challenger explosion became a significant event in American history and its cultural impact:

1. National Tragedy:

The Challenger disaster was a national tragedy that deeply affected the American public. The loss of seven astronauts, including Christa McAuliffe, the first civilian teacher selected to fly in space, struck a chord with people of all ages. The accident captured the collective grief and empathy of the nation, as people mourned the loss of the brave astronauts who had embarked on a historic mission. ˙

2. Media Coverage and Public Witness:

The launch of the Space Shuttle Challenger was extensively covered by the media. The shocking images of the shuttle's explosion were replayed repeatedly on television, leaving a lasting imprint on the public's memory. The incident became a shared experience for many Americans, who witnessed the tragedy unfold in real-time.

3. Impact on Space Exploration:

The Challenger explosion had significant implications for space exploration. It resulted in a temporary suspension of the Space Shuttle program, triggering a thorough investigation and reevaluation of safety protocols. The disaster prompted a renewed commitment to crew safety and risk management in future space missions.

4. Cultural Iconography:

The Challenger explosion became an indelible cultural iconography representing both human exploration and its inherent risks. The image of the Space Shuttle breaking apart seconds after liftoff became a symbol of tragedy and the price of pushing the boundaries of human knowledge.

5. Impact on Education:

Christa McAuliffe's presence on the Challenger mission as the "Teacher in Space" had a profound impact on education in the United States. Her enthusiasm for teaching and dedication to promoting STEM education left a lasting legacy. The tragedy prompted educators and policymakers to reinforce the importance of science education and space exploration in schools.

6. Challenger Crew as Heroes:

CHALLENGER: TRAGEDY AND TRIUMPH - UNRAVELING THE SPACE SHUTTLE CHALLENGER EXPLOSION

The Challenger crew members were remembered and revered as heroes who exemplified courage and dedication to exploration. Their sacrifice became a source of inspiration for generations to come, encouraging individuals to pursue careers in space science, engineering, and related fields.

7. Investigative Commissions:

The Presidential Commission on the Space Shuttle Challenger Accident, known as the Rogers Commission, conducted an in-depth investigation into the disaster. The commission's findings and recommendations had a lasting impact on space exploration, influencing safety practices and organizational culture within NASA.

8. Public Perception of Space Exploration:

The Challenger explosion had a profound impact on public perception of space exploration. It brought to the forefront the inherent risks involved in human spaceflight and raised questions about the feasibility and necessity of manned missions. It also prompted debates about the role of space exploration in the nation's priorities.

The Challenger explosion became a significant event in American history due to its national impact, media coverage, and cultural resonance. It transformed space exploration, leaving a legacy of improved safety practices and a renewed commitment to the pursuit of knowledge. The memory of the Challenger crew continues to inspire future generations, fostering a spirit of exploration, scientific curiosity, and respect for the bravery and dedication of those who venture into the unknown to expand the horizons of human understanding.

CHALLENGER: TRAGEDY AND TRIUMPH - UNRAVELING THE SPACE SHUTTLE CHALLENGER EXPLOSION

Chapter 18: Challenger's Controversies and Conspiracy Theories

———

The Challenger explosion, being a significant and highly publicized event, has unfortunately been subject to various controversies and conspiracy theories over the years. It's essential to approach these claims with critical thinking and rely on factual evidence and expert analysis. Here are some of the controversies and conspiracy theories that have emerged surrounding the Challenger explosion:

1. O-Ring Controversy:

One of the main controversies surrounding the Challenger disaster revolves around the O-rings, which were part of the solid rocket boosters (SRBs). The investigation revealed that the O-rings failed due to the cold weather on the day of the launch. Some have questioned the engineering decision to launch in such weather conditions and whether NASA could have prevented the disaster by postponing the launch.

2. Cover-Up Allegations:

Some conspiracy theories suggest that there was a cover-up by NASA and government officials to hide information or mislead the public about the true cause of the Challenger explosion. These claims assert that the disaster was not a result of the O-ring failure but was caused by other factors, and the truth was intentionally suppressed.

3. Sabotage Theories:

Conspiracy theories have emerged, suggesting that the Challenger explosion was not an accident but rather an act of sabotage or deliberate destruction. These claims lack credible evidence and have been widely debunked by experts.

4. Political and Budgetary Motivations:

Some critics have pointed to political or budgetary motivations, suggesting that the pressure to maintain the Shuttle program's schedule and secure funding for NASA played a role in the decision to proceed with the launch despite safety concerns.

5. Allegations of Psychological Testing:

There have been claims that Christa McAuliffe, the "Teacher in Space" and one of the Challenger crew members, was chosen for the mission not solely for her educational background but also because she underwent psychological testing or had specific personality traits.

It is crucial to note that the official investigation into the Challenger disaster, led by the Presidential Commission on the Space Shuttle Challenger Accident (Rogers Commission), conducted a thorough examination of the evidence and concluded that the O-ring failure was the primary cause of the accident. The commission's report, released in June 1986, remains the most authoritative account of the disaster.

CHALLENGER: TRAGEDY AND TRIUMPH - UNRAVELING THE SPACE SHUTTLE CHALLENGER EXPLOSION

In the face of controversies and conspiracy theories, the scientific and engineering community's consensus supports the conclusions of the Rogers Commission. It is essential to rely on credible sources and expert analysis when discussing the Challenger explosion and to remember the bravery and sacrifice of the seven astronauts who lost their lives in pursuit of exploration. Focusing on factual evidence and learning from the lessons of the Challenger disaster can contribute to improving safety and advancing space exploration endeavors.

1. O-Ring Controversy:

The O-ring failure has been extensively investigated and is well-documented as the primary cause of the Challenger explosion. The Rogers Commission, consisting of experts from various fields, conducted a thorough examination of the accident. They concluded that the O-rings, which were designed to seal joints in the solid rocket boosters, did not function properly due to the cold weather on the day of the launch. This failure allowed hot gases to escape, leading to the catastrophic explosion.

Subsequent studies and investigations by experts in the aerospace and engineering fields have consistently supported the commission's findings. The O-ring issue was a known concern within NASA and the contractor responsible for the boosters. Despite warnings and objections from engineers, the launch proceeded, resulting in the tragic outcome.

2. Cover-Up Allegations:

There is no credible evidence to support the claim of a cover-up by NASA or government officials regarding the Challenger disaster. The investigation conducted by the Rogers Commission was thorough and transparent, and its findings were made public. The commission's report provided a detailed analysis of the technical and organizational factors that contributed to the accident.

The process of investigating high-profile accidents like the Challenger disaster involves multiple independent experts, government agencies, and industry professionals. Concealing or misrepresenting crucial information in such investigations would be highly improbable and ethically unacceptable. The evidence strongly supports the conclusion that the Challenger explosion was the result of the O-ring failure and not the outcome of any deliberate cover-up.

3. Sabotage Theories:

Claims of sabotage or deliberate destruction of the Challenger spacecraft lack credible evidence and have been widely debunked by experts. The physical evidence, including the remains of the spacecraft and the booster rockets recovered from the ocean, points to a catastrophic failure in the solid rocket boosters due to O-ring failure.

Accidents involving complex systems like space shuttles typically result from a combination of technical, engineering, and organizational factors, rather than deliberate sabotage. Investigations into accidents of this magnitude are carried out meticulously by teams of experts, with a focus on uncovering the root causes to prevent future occurrences.

4. Political and Budgetary Motivations:

While budgetary constraints and the pressure to maintain the Shuttle program's schedule were factors in NASA's decision-making process, there is no credible evidence to suggest that they played a direct role in the Challenger disaster. The primary cause of the accident, as established by the Rogers Commission and supported by experts, was the O-ring failure, exacerbated by management decisions and communication breakdowns.

5. Allegations of Psychological Testing:

The claim that Christa McAuliffe was selected for the Challenger mission based on psychological testing or specific personality traits lacks credible evidence. The Teacher in Space Project, which aimed to involve a teacher in a spaceflight, went through a rigorous selection process. Christa McAuliffe, a dedicated and enthusiastic educator, was selected for her passion for teaching and her ability to communicate the excitement of space exploration to students.

The available evidence and expert opinions overwhelmingly support the findings of the Rogers Commission regarding the cause of the Challenger explosion. The O-ring failure due to cold weather conditions remains the primary reason for the tragic accident. Claims of cover-ups, sabotage, and other conspiracy theories lack credible evidence and are not supported by the scientific and engineering community. The focus should remain on the lessons learned from the disaster to enhance safety and advance space exploration for the benefit of humanity.

CHALLENGER: TRAGEDY AND TRIUMPH - UNRAVELING THE SPACE SHUTTLE CHALLENGER EXPLOSION

Chapter 19: The Challenger Memory in Pop Culture

The Challenger disaster has been portrayed and remembered in various books, movies, documentaries, and other media formats. As a significant event in American history, it continues to serve as a source of inspiration, exploration, and reflection. Here's an examination of how the Challenger disaster has been portrayed and remembered in different forms of media:

1. Books:

Numerous books have been written about the Challenger disaster, offering detailed accounts of the events leading up to the tragedy, the investigation, and its aftermath. Some books focus on the technical aspects, exploring the engineering and organizational factors that contributed to the accident. Others delve into the personal stories of the crew members and their families, providing a human perspective on the impact of the disaster.

2. Documentaries:

Several documentaries have been produced to chronicle the Challenger disaster, offering in-depth analysis and insights into the causes of the accident. These documentaries often feature interviews with experts, engineers, astronauts, and family members of the Challenger crew, providing a comprehensive and emotional look at the events surrounding the tragedy.

3. Feature Films and Dramatizations:

The Challenger disaster has been the subject of several feature films and dramatizations. Some movies have portrayed the events leading up to the launch, the decision-making process, and the emotional toll on those involved. These films aim to educate viewers about the complexity of space missions, the importance of safety, and the consequences of taking risks.

4. Educational and Historical Programs:

The Challenger disaster is often featured in educational and historical programs, especially those focusing on space exploration, engineering ethics, and safety protocols. These programs use archival footage, interviews, and expert analysis to present a comprehensive understanding of the events and the lessons learned.

5. Tributes and Memorials:

Various media forms, including documentaries and books, have paid tribute to the Challenger crew and their legacy. These tributes celebrate the bravery and dedication of the astronauts, emphasizing their contributions to space exploration and education. Memorials and anniversary events also remember the Challenger crew, reminding the public of the risks and rewards of human spaceflight.

6. Impact on Space Policy and Public Perception:

CHALLENGER: TRAGEDY AND TRIUMPH - UNRAVELING THE SPACE SHUTTLE CHALLENGER EXPLOSION

The Challenger disaster has also been discussed in the context of space policy and public perception of space exploration. Media coverage has explored the lessons learned from the tragedy, how it influenced space policy, and the ongoing efforts to prioritize safety in human spaceflight.

7. Educational Initiatives and Outreach:

In addition to traditional media, the Challenger disaster has influenced educational initiatives and outreach programs. It continues to inspire educators and students to pursue STEM education and careers in space-related fields, fostering a new generation of scientists, engineers, and explorers.

The Challenger disaster has been portrayed and remembered in various media forms, serving as a reminder of the sacrifices and challenges of space exploration. Through books, movies, documentaries, and educational programs, the tragedy continues to influence the public's understanding of space exploration, safety, and the human spirit of exploration. It serves as a cautionary tale and a source of inspiration, motivating the space community to learn from past mistakes and strive for safer, more ambitious space missions in the future.

OLIVER LANCASTER

Chapter 20: Looking Back - Reflections and Insights

―――

The Challenger explosion remains an indelible and tragic event in the history of space exploration, leaving a lasting impact on NASA, the aerospace industry, and the public. The legacy of the Challenger disaster is multifaceted, encompassing profound changes in safety culture, engineering practices, and public perception of space exploration. Despite the heartbreak and loss, the lessons learned from the tragedy have helped shape a safer and more resilient future for human spaceflight.

The primary lesson from the Challenger explosion was the imperative of prioritizing crew safety above all else. The disaster prompted a fundamental transformation in safety culture within NASA and the aerospace industry, emphasizing the need for meticulous risk assessments, open communication, and continuous learning. Space agencies and private companies now incorporate comprehensive safety measures and redundant systems to protect astronauts and spacecraft during space missions.

The Challenger disaster also sparked advancements in spacecraft design, technology, and engineering. The lessons learned have contributed to the development of new, safer spacecraft, like the Orion crew capsule and the Space Launch System (SLS), enabling deep space exploration missions to the Moon, Mars, and beyond.

Beyond its technical impact, the Challenger explosion has left a lasting cultural legacy. The memory of the brave Challenger crew members continues to inspire future generations, fostering a spirit of exploration and scientific curiosity. Their dedication to education, exemplified by Christa McAuliffe as the "Teacher in Space," has had a profound influence on STEM education, motivating students to pursue careers in science, technology, engineering, and mathematics.

The Challenger explosion has also shaped public perception of space exploration. It serves as a reminder of the risks and challenges inherent in human spaceflight, encouraging a cautious and calculated approach to exploration. However, it also underscores the resilience and determination of the human spirit in the face of adversity, motivating the space community to continue pushing the boundaries of human knowledge and expanding our presence in the cosmos.

The Challenger explosion's legacy is a profound one, leaving an enduring impact on space exploration and the collective consciousness of humanity. The tragedy serves as a constant reminder of the importance of safety, the need for continuous learning and improvement, and the dedication of those who dare to venture beyond Earth's boundaries. As we honor the memory of the Challenger crew and continue to explore the cosmos, their sacrifice remains a poignant reminder of the challenges and rewards of human spaceflight, fueling our passion for exploration and inspiring future generations to reach for the stars.

CHALLENGER: TRAGEDY AND TRIUMPH - UNRAVELING THE SPACE SHUTTLE CHALLENGER EXPLOSION

The ongoing quest for space exploration is a testament to humanity's enduring spirit to push the boundaries of knowledge and discovery. Since the early days of space travel, the dream of exploring the cosmos has captivated the imagination of people worldwide. This unyielding curiosity and adventurous spirit have led to remarkable achievements and opened new frontiers in human understanding.

Exploration has always been an integral part of human history. From ancient seafarers navigating uncharted waters to modern-day astronauts venturing into the vastness of space, the quest for discovery has shaped the evolution of civilization. The exploration of space, in particular, has transcended national borders and united humanity in its shared pursuit of knowledge and understanding.

Throughout the space age, humanity has achieved awe-inspiring milestones. The first human in space, Yuri Gagarin, made a historic orbit of the Earth in 1961. The Apollo program enabled humans to set foot on the Moon, with Neil Armstrong's famous words, "That's one small step for [a] man, one giant leap for mankind," echoing across history. Robotic missions have explored distant planets, moons, and asteroids, providing invaluable insights into the mysteries of our solar system and beyond.

The International Space Station (ISS) stands as a testament to international collaboration in space exploration. This orbiting laboratory serves as a microcosm of global unity, where astronauts from various countries live and work together in pursuit of scientific research and technological advancements.

The ISS showcases humanity's capacity for cooperation in the face of complex challenges.

As we move forward, the vision of exploring deeper into space continues to inspire scientists, engineers, and visionaries. Mars remains a prime target for future crewed missions, with efforts to study the Red Planet's surface and potential for human settlement well underway. Technologies for interplanetary travel and long-duration missions are constantly evolving, bringing us closer to the possibility of becoming a multi-planetary species.

Beyond Mars, humanity's quest for space exploration extends to exoplanets and distant galaxies. Observatories like the Hubble Space Telescope and future space telescopes offer glimpses of worlds beyond our solar system, fueling the excitement of discovering potentially habitable planets and life beyond Earth.

However, the pursuit of space exploration is not without challenges. It requires a delicate balance between ambition and caution. Ensuring the safety and well-being of astronauts, developing sustainable space technologies, and protecting the environments we explore are crucial considerations. Responsible exploration is essential to preserve the pristine nature of celestial bodies for future generations.

The ongoing quest for space exploration is a manifestation of the human spirit to dream, discover, and transcend our limitations. It embodies the belief that the pursuit of knowledge and understanding is a noble endeavor that enriches our collective consciousness. As we continue to explore the cosmos, humanity stands on the cusp of unprecedented discoveries, which may redefine our place in the universe and ignite our imaginations for generations to come.

Ultimately, the enduring spirit of space exploration is a testament to the resilience, ingenuity, and cooperation of humanity. It reminds us that no challenge is too great, no horizon too distant, and that the desire to explore is an integral part of what makes us human. The journey into space embodies hope, wonder, and the unshakable belief that there is more to be discovered and explored, encouraging us to reach ever further into the cosmos and beyond.

Timeline of Key Events Leading up to and Following the Space Shuttle Challenger Explosion:

January 28, 1986:

- 11:38 a.m. EST: Space Shuttle Challenger (mission STS-51-L) lifts off from Kennedy Space Center in Florida with seven crew members aboard.

- 73 seconds after liftoff: The shuttle explodes in mid-air, disintegrating into flames. All seven astronauts onboard perish in the tragedy.

Leading Up to the Explosion:

- 1980s: The Space Shuttle program aims to make space travel more routine and cost-effective, conducting numerous successful missions before the Challenger disaster.

- January 23, 1986: The Challenger's initial launch is delayed due to technical issues, including problems with the solid rocket boosters' O-rings in cold weather.

- January 27, 1986: Engineers at Morton Thiokol, the contractor responsible for the solid rocket boosters, express concerns about the O-rings' performance in freezing temperatures.

After the Explosion:

- January 28, 1986: President Ronald Reagan addresses the nation and pays tribute to the fallen astronauts.

- February 3, 1986: The Presidential Commission on the Space Shuttle Challenger Accident (Rogers Commission) is established to investigate the disaster.

- June 6, 1986: The Rogers Commission presents its final report, citing the O-ring failure as the cause of the explosion and recommending significant changes to the Shuttle program.

Safety Reforms and Improvements:

- 1988: NASA resumes shuttle flights after implementing numerous safety reforms, including redesigning the solid rocket boosters, improving communication channels, and enhancing safety protocols.

- 1990: The Hubble Space Telescope is deployed, a testament to the continued exploration of space despite the Challenger disaster.

CHALLENGER: TRAGEDY AND TRIUMPH - UNRAVELING THE SPACE SHUTTLE CHALLENGER EXPLOSION

- 1993: Space Shuttle Endeavour is launched for the first time, marking a new era in the Space Shuttle program following the Challenger tragedy.

Challenger's Legacy and Impact:

- The Challenger Learning Center program is established in 1986 to honor the crew's commitment to education, providing hands-on science and space education experiences for students worldwide.

- The Challenger Center for Space Science Education is founded in 1987, supporting STEM education and outreach initiatives in memory of the Challenger crew.

- The legacy of the Challenger crew is commemorated through numerous memorials, documentaries, books, and educational programs, emphasizing the importance of safety, dedication, and exploration in space missions.

Space Exploration Beyond the Shuttle Program:

- 1998: The International Space Station (ISS) begins its assembly in orbit, signifying a collaborative international effort in space exploration.

- 2003: Space Shuttle Columbia disintegrates upon reentry, leading to another investigation into the Space Shuttle program's safety and reinforcing the lessons learned from the Challenger disaster.

- 2011: The Space Shuttle program is retired, marking the end of an era in human spaceflight, but paving the way for future exploration missions.

Present and Future Endeavors:

- 2021 and beyond: Continued robotic exploration missions to Mars and beyond, advancements in private spaceflight, and ambitious plans for crewed missions to the Moon and Mars, symbolize the enduring spirit of space exploration that perseveres despite challenges.

The Challenger disaster's impact on space exploration has been profound, shaping safety protocols, educational initiatives, and fostering a commitment to advancing knowledge and discovery in the cosmos. The legacy of the Challenger crew endures, inspiring future generations to reach for the stars and explore the unknown.

CHALLENGER: TRAGEDY AND TRIUMPH - UNRAVELING THE SPACE SHUTTLE CHALLENGER EXPLOSION

Sign up to my free newsletter to get updates on new releases, FREE teaser chapters to upcoming releases and FREE digital short stories.

Or visit https://tinyurl.com/olanc

I never spam and you can unsubscribe at any time.

Don't miss out!

Visit the website below and you can sign up to receive emails whenever Oliver Lancaster publishes a new book. There's no charge and no obligation.

https://books2read.com/r/B-A-UNEZ-TIIMC

BOOKS 2 READ

Connecting independent readers to independent writers.

Also by Oliver Lancaster

Chernobyl: Unveiling the tragedy. A Comprehensive Account
of the Nuclear Disaster
The Bhopal Gas Tragedy: Unraveling the Catastrophe of 1984
The Deepwater Horizon Oil Spill of 2010: A Disaster
Unveiled
Fukushima Fallout: Unveiling the Truth behind the 2011
Nuclear Disaster
Minamata Disease: Poisoned Waters and the Battle for Justice
(1932-1968)
Evil Women: Unmasking History's Most Notorious Women
Bundy The Dark Chronicles: America's Infamous Serial Killer
Dahmer The Dark Chronicles: America's Infamous
Milwaukee Cannibal
Zodiac The Dark Chronicles: America's Infamous Cryptic
Killer
Bigfoot: The Comprehensive Investigation into the Elusive
Legend
Chasing Legends: The Truth behind the Chupacabra
Chasing Legends: The Truth behind the Loch Ness Monster
Aokigahara Forest: The Heartbreaking Secrets of Japan's
Suicide Forest
The Amityville House: The Haunting Secrets of America's
Most Infamous Residence

Watch for more at https://tinyurl.com/olanc.

About the Author

Oliver Lancaster possesses an enchanting charm that effortlessly draws readers into the depths of his literary world. With an insatiable curiosity for the unexplained, he skillfully weaves tales of crime, conspiracy, mystery and the unknown, leaving readers on the edge of their seats.

Nestled away in the seclusion of his garden shed, Oliver finds solace and inspiration in the tranquility of nature. Surrounded by greenery and fragrant blooms, he dives into a realm of imagination, unearthing secrets that lie hidden within his mind.

Accompanying Oliver on his literary ventures is his faithful ginger cat named Italics. With his mesmerizing gaze and mysterious mannerisms, Italics adds an air of intrigue to Oliver's writing process, often curling up on a cushioned chair nearby, watching as words flow effortlessly from his human companion's pen.

When not engrossed in his craft, Oliver indulges in the gentle warmth of his garden with a glass of red wine.

Prepare to be spellbound as you delve into the pages of Oliver Lancaster's novels, for he is a master of the eerie, a weaver of secrets, and an unrivaled guide through the labyrinthine corridors of the human psyche.

Sign up to a free newsletter to get updates on new releases, FREE teaser chapters to upcoming releases and FREE digital short stories.

Read more at https://tinyurl.com/olanc.